Family Farm Fun
A Satirical Activity & Game Book about the Hazards of Industrial Farming
Doug Baird

ISBN 978-0-9898608-3-3

Preface

Early Sunday morning, June 4, 2017, I was sicker than I had ever been before. Too sick to even bend over, as I vomited all over the toilet, myself, and the bathroom floor — and I didn't even care.

This was the aftermath of being engulfed in a cloud of Roundup from a giant agricultural sprayer while I was mowing my lawn the previous afternoon.

The resulting NYSDEC investigation ignored weather records showing wind speeds in excess of 20 mph, merely stating that the applicator "disputed the wind speed," and characterized me as someone who "expressed views that was politically active against farming" in spite of my having rented that field to a farmer for the last 25 years.

Helena Chemical was only issued a warning for "application of pesticide to non-target area" and the report ended with an unequivocal "Case closed!"

This incident motivated me to write *You Know You Live near a Factory Farm When Your Kids Go Fishing with a Pool Skimmer* — a picture book with large print and cautionary captions. The *Family Farm Fun* book is the second in the Factory Farm series.

Mine is not an isolated case. Just a hint of interest will elicit local accounts of herbicide sprayed onto wash hanging out to dry and families decimated by cancer.

The rural community is nothing like it was 25 years ago. Many hard-working rural people have died, moved, or been forced out, wells and streams have been polluted from agricultural runoff and manure spills, and rich corporate agribusinesses have taken over. The drugs, crime, and poverty of this new "ag ghetto" are contained by a sheriff's department who only come when they're called, and leave everything unchanged.

These days I think of my health differently . . . I look for signs of cancer.

I would like to thank the members of the Lansing, NY Writers Group for their many helpful comments and suggestions, and a special thank you to the creators of the distinctive fonts used throughout this book for the value they added to its pages.

Doug Baird

Lansing, N.Y., *July 13, 2019*

Contents

Games

It's Ag-endemic

Mazes

Poems

Songs

Escape from the Factory Farm

You need one die from a set of dice for this game, or you can make a spinner as shown below.
Start by throwing or spinning a six, and continue your escape until you reach the end. The first one to reach the end wins.

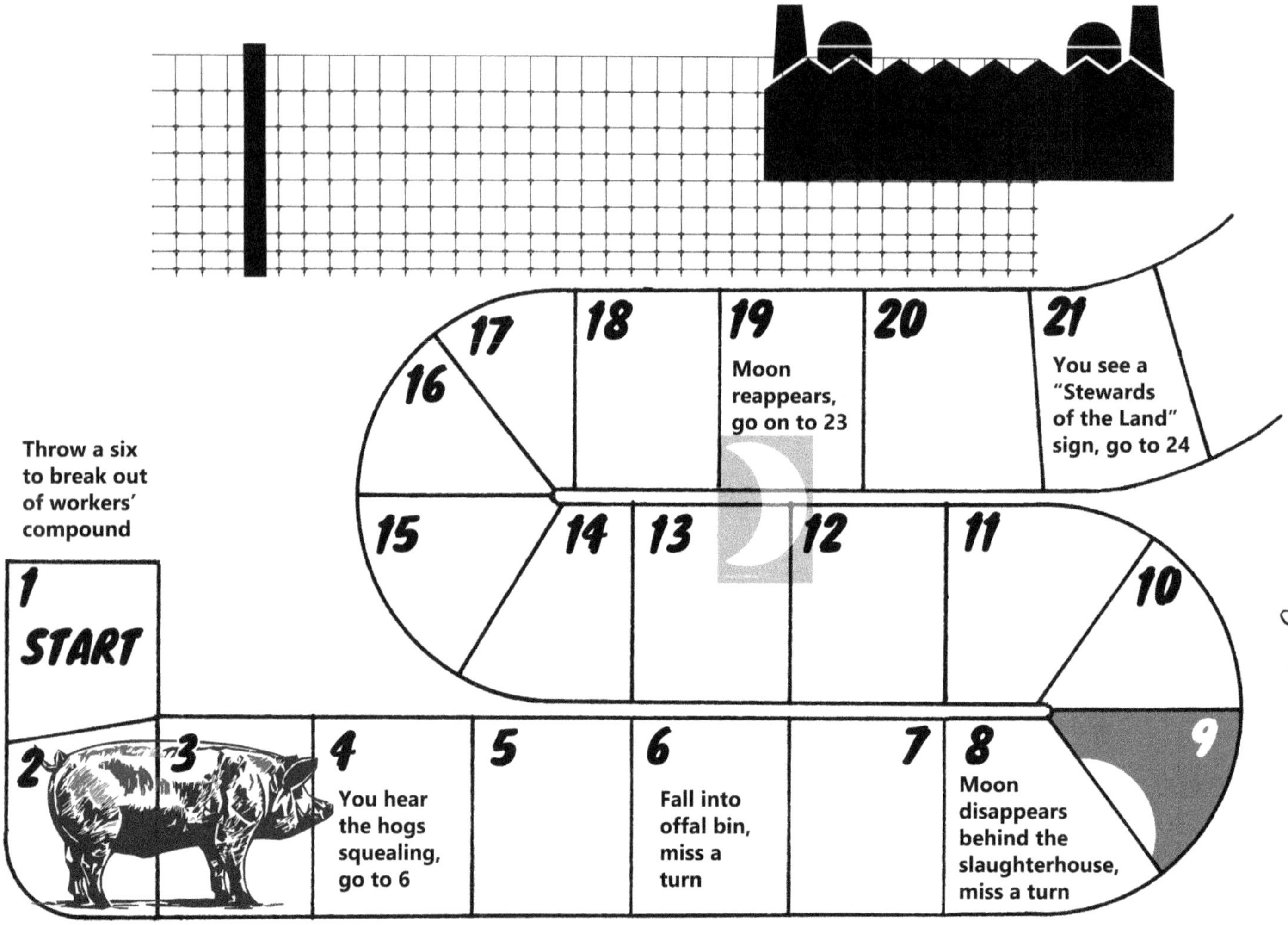
Throw a six to break out of workers' compound
1 START
2
3
4 You hear the hogs squealing, go to 6
5
6 Fall into offal bin, miss a turn
7
8 Moon disappears behind the slaughterhouse, miss a turn
9
10
11
12
13
14
15
16
17
18
19 Moon reappears, go on to 23
20
21 You see a "Stewards of the Land" sign, go to 24

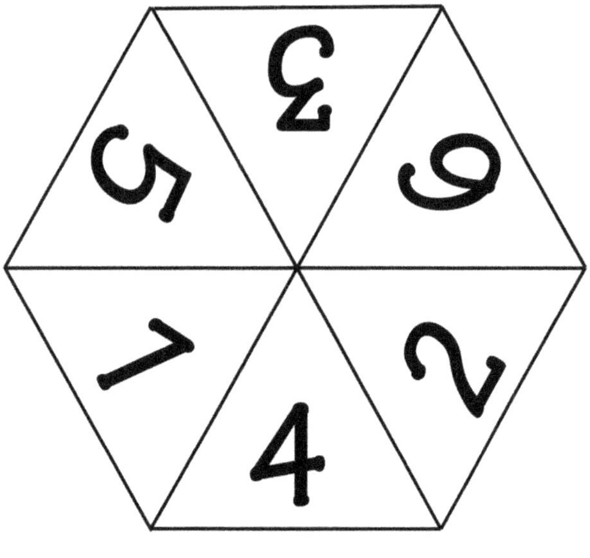
3
5
6
1
2
4

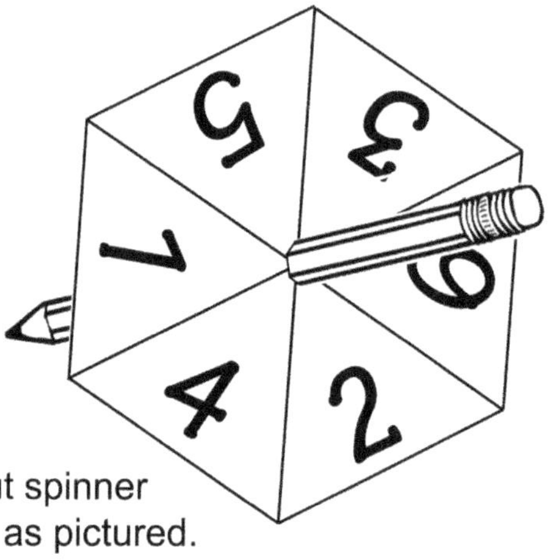
5
3
1
6
4
2

Copy or cut out spinner and assemble as pictured.

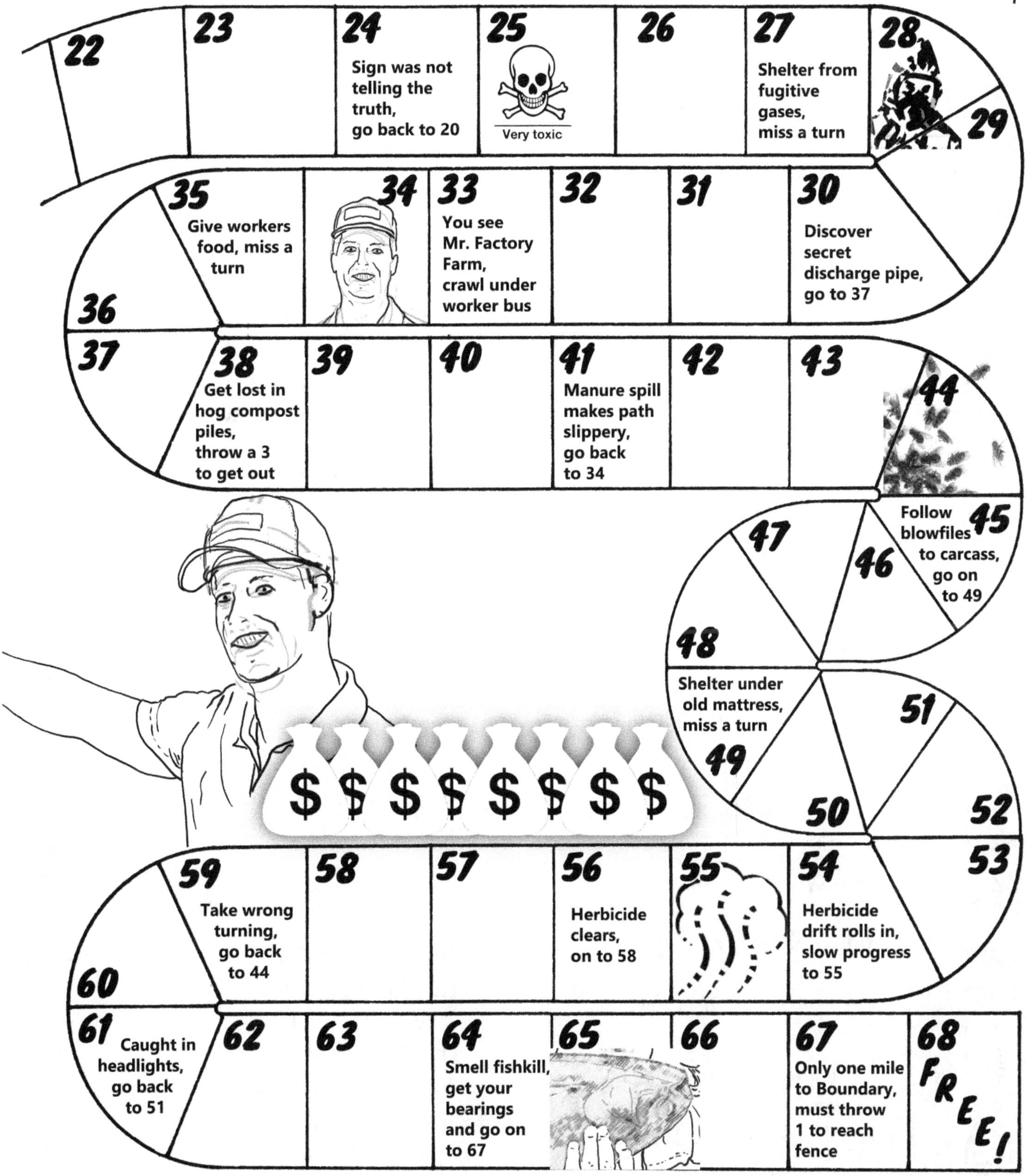
22
23
24
Sign was not telling the truth, go back to 20
25
Very toxic
26
27
Shelter from fugitive gases, miss a turn
28
29
30
Discover secret discharge pipe, go to 37
31
32
33
You see Mr. Factory Farm, crawl under worker bus
34
35
Give workers food, miss a turn
36
37
38
Get lost in hog compost piles, throw a 3 to get out
39
40
41
Manure spill makes path slippery, go back to 34
42
43
44
45
Follow blowfiles to carcass, go on to 49
46
47
48
49
Shelter under old mattress, miss a turn
50
51
52
53
54
Herbicide drift rolls in, slow progress to 55
55
56
Herbicide clears, on to 58
57
58
59
Take wrong turning, go back to 44
60
61
Caught in headlights, go back to 51
62
63
64
Smell fishkill, get your bearings and go on to 67
65
66
67
Only one mile to Boundary, must throw 1 to reach fence
68
FREE!
$ $ $ $ $ $ $ $

Q; When is it unlucky to see a factory farm? A: When it's out of your bedroom window.

New School Nursery Rhymes

Farming – It's not your "Old MacDonald" anymore

Industrial Farming has more in common with mining and manufacturing than traditional farming. To honor this change from pastoral to pharmaceutical and bucolic to bureaucratic – four traditional nursery rhymes have been updated to reflect our modern factory farm ethos.

Use the titles below (or choose your favorites) and create your own new millennium rhymes.

Itsy-bitsy pathogen

Polly Put the Particle Mask On

To market, to market to buy some pink slime

This is the Dead Zone that Jack built

A-Tisket, A-Hazmat

Spill Spill Go Away

Sing a Song of Spillage

Three Blind Regulators

I'm a Little Cesspit

The Wheels on the Sprayer

Poop Goes the CAFO

Mary's Lamb Revised

Mary had a little lamb
Its dad was GMO
And when it ate the family cat
The kid was next to go.

Hey Diddle Diddle Revised

Hey, "Piddle Piddle"
The worker was little
The cow fell in the lagoon
The rich farmer laughed
To see that shit
And the fish kill stank
all afternoon.

Mary, Mary, Quite Contrary Revised

Mary, Mary, giant dairy,
How do your profits grow?
With Ag cartels and
corporate shells,
And a Congress that
never says "no."

Rock-a-bye Baby Revised

Rock-a-blue baby
The flies never stop
When the wind blows
The drift is a shock
When lagoons break
Manure will spill
And nitrates for baby
Who lies there so still.

Knock, knock — Who's there? — Pink slime — Pink slime who? — Shhhh . . .

Life Saving -
Child Safety Tips

You can't always be near your child. How can you make sure they are safe? Teach your children:

FARM HARM

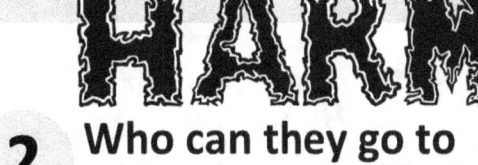

1 What do you mean by farm harm?

Modern farming practices produce food filled with hormones and antibiotics, and pollute the air, land and water with toxic substances — putting your child at risk.

2 Who can they go to when they need help?

X Local Authorities
X DEC
X USDA

3 Possible dangerous situations, e.g.:

An authority figure tells your child: milk from cows treated with bovine growth hormones is safe to drink.

Children should only be given what has been proven safe, *not* what has not yet been proven unsafe.

Your child wishes to swim at the local swimming hole you used as a kid.

Fecal matter, coliform bacteria and antibiotics from factory farming may be present. All water around farms should be tested before use.

4 What to teach your child:

Do not touch anything around a farm - it may be bad.

Talk with mummy/daddy about what you eat and drink.

Scream or shout "NO" when they are near farm harm.

NO!

Parents

Agricultural laws are intended to protect the profits of farmers, not the health of your children.
Teach them about Farm Harm.

"Because Industrial Farming changes everything."

HYDROGEN
H2S
SULFIDE

The Widow Maker

NUTRIENT
POLLUTION

Dead Zone

ManureSpillPublicOutcryCommitteeReportVoluntaryGuidelines
REGULATIONS

The Runaround

BOVINE
GROWTH
HORMONE

Tit-for-Tat

Q: How many thriving rural communities have factory farms? A: None.

The Industrial Agriculture Anthem

(To the tune of "America the Beautiful")

O beautiful bodacious lies
For ample yields of gain
For purposes we must disguise
Those riches to obtain!
America! America!
I wave the flag for me
So that my good [It's understood]
Is worth much more than thee!

O beautiful for that defeat
Of those that I transgress
What matters if I lie and cheat
My truth is my success!
America! America!
Just pass those nuisance laws
Give me control to reach my goal
Expansion without pause!

O beautiful for things unproved
Cut nature like a knife
And those who lands and waters loved
Were merely fools for life!
America! America!
May gold my god define
So my excess be never less
And every grain be mine!

O beautiful for worldly dream
That dies beyond my years
With politicians close I scheme
Unmoved by human tears!
America! America!
You'll buy your food from me
And crown my good [as well you should]
From sea to dying sea!

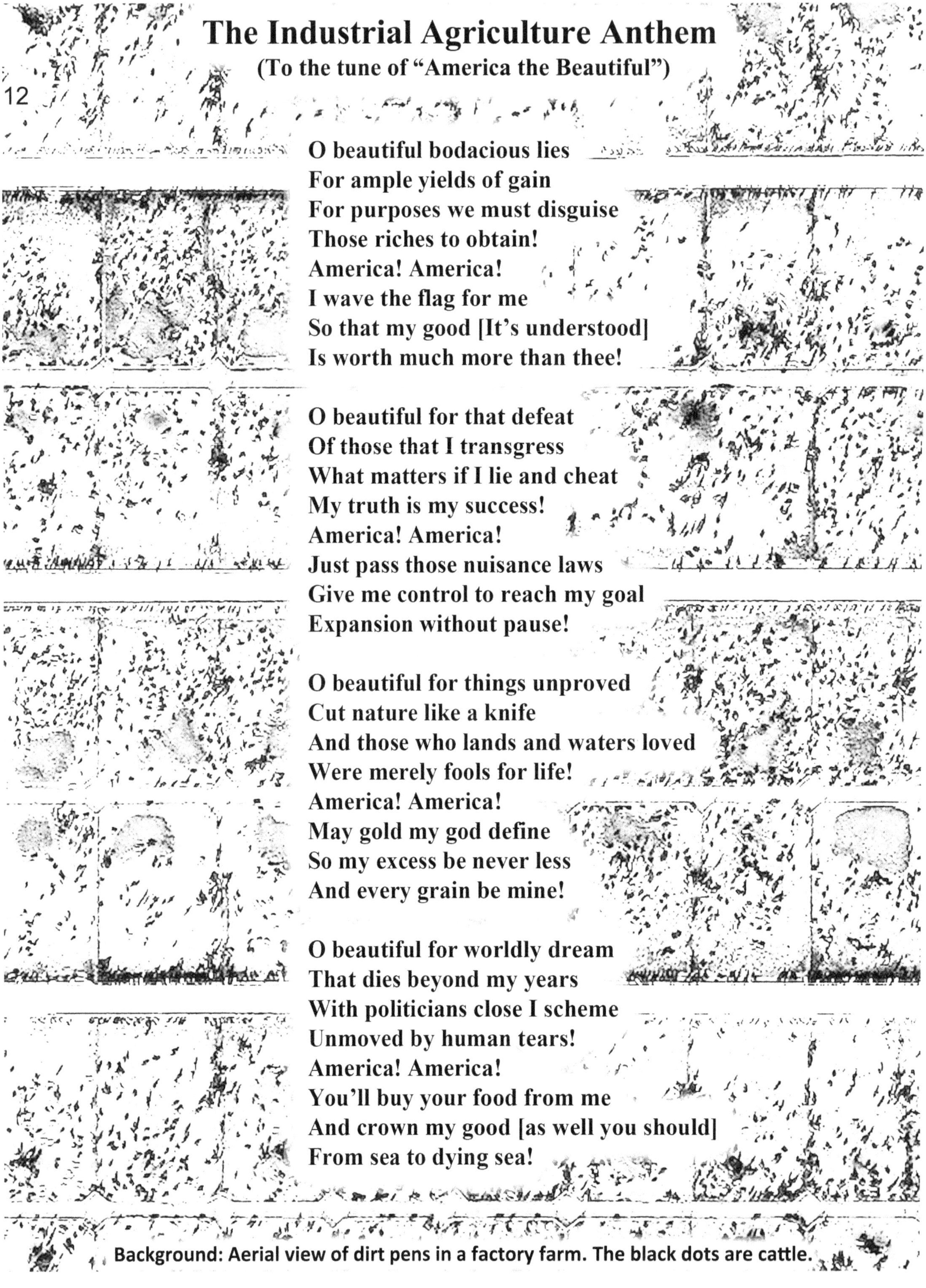

Background: Aerial view of dirt pens in a factory farm. The black dots are cattle.

Fish Kill Crunch

Make fish crackers and cookies using fish shaped-cutters. [Be sure to poke a hole through the middle of each fish before baking.]

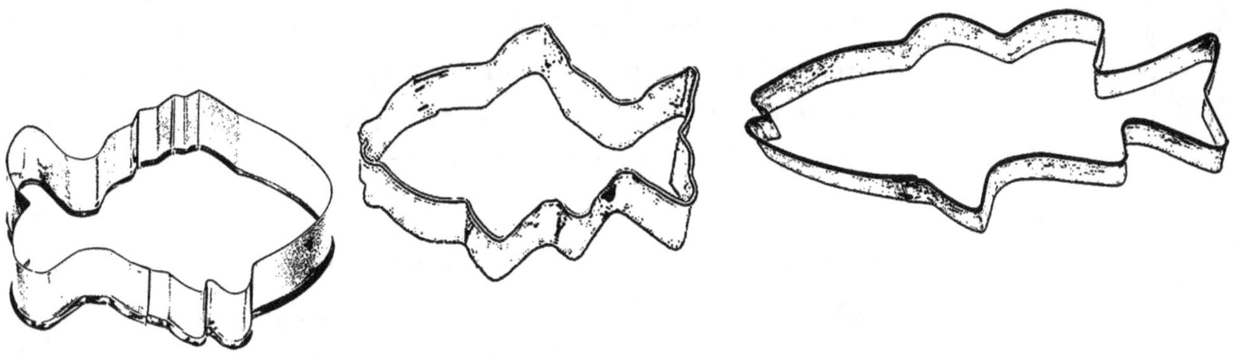

Thread equal amounts of the baked fish kill onto two long pieces of string representing stretches of a river, and have adults hold the strings up at a child's height.

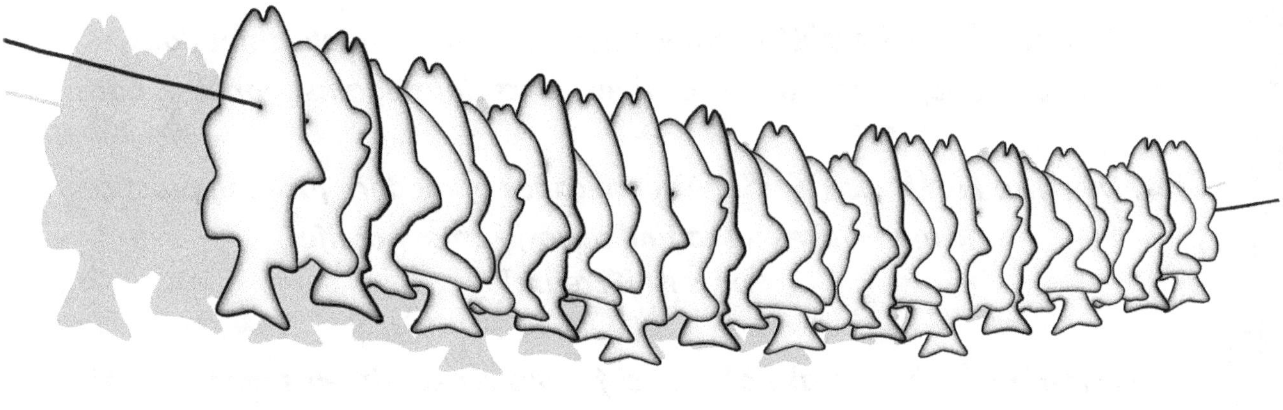

Divide the children into two teams. With their hands behind their backs, they've got to eat all the fish until the string is empty.

The first team to clean up their stretch of the river is the winner. Until the next time.

Monster Movies

Creature from the Manure Lagoon – *When antibiotic-resistant bacteria start eating flesh, heroic health care professionals seal off the rural community to prevent its spread.*

It Came from Beneath the C – *A college's underground Ag lab produces a genetically modified insect that becomes uncontrollable, devouring the town.*

War of the World – *The epic spectacle of rich corporate farms using herbicides, genetically modified organisms, and Right to Farm laws to batter a world of nuisances into submission.*

Attack of the 50,000 Animal Farm – *An unstoppable giant dairy destroys air, land, and water as local residents fight against hopeless odds.*

The Day the Earth was Still – *After decades of modern farming practices, one man wakes up to find a world empty of life . . . or is it! [Scream]*

Farmzilla – *In this remake of the Sci-fi classic, experiments in industrial farming have created a 8,000 square mile monster of destruction off the Gulf Coast. Can it be stopped in time!?*

Q: What's the scariest word a factory farmer can hear? A: Regulate

Spread It!

Nutrient Plan Solitaire

Spread It! is a "simple addition" solitaire game using either a standard 52 card or Factory Farm deck. The object is to remove pairs of cards totaling 13.

The game is begun by dealing 28 cards, face up, to the "Nutrient Plan" in the form of a triangle. The layout will look something like this:

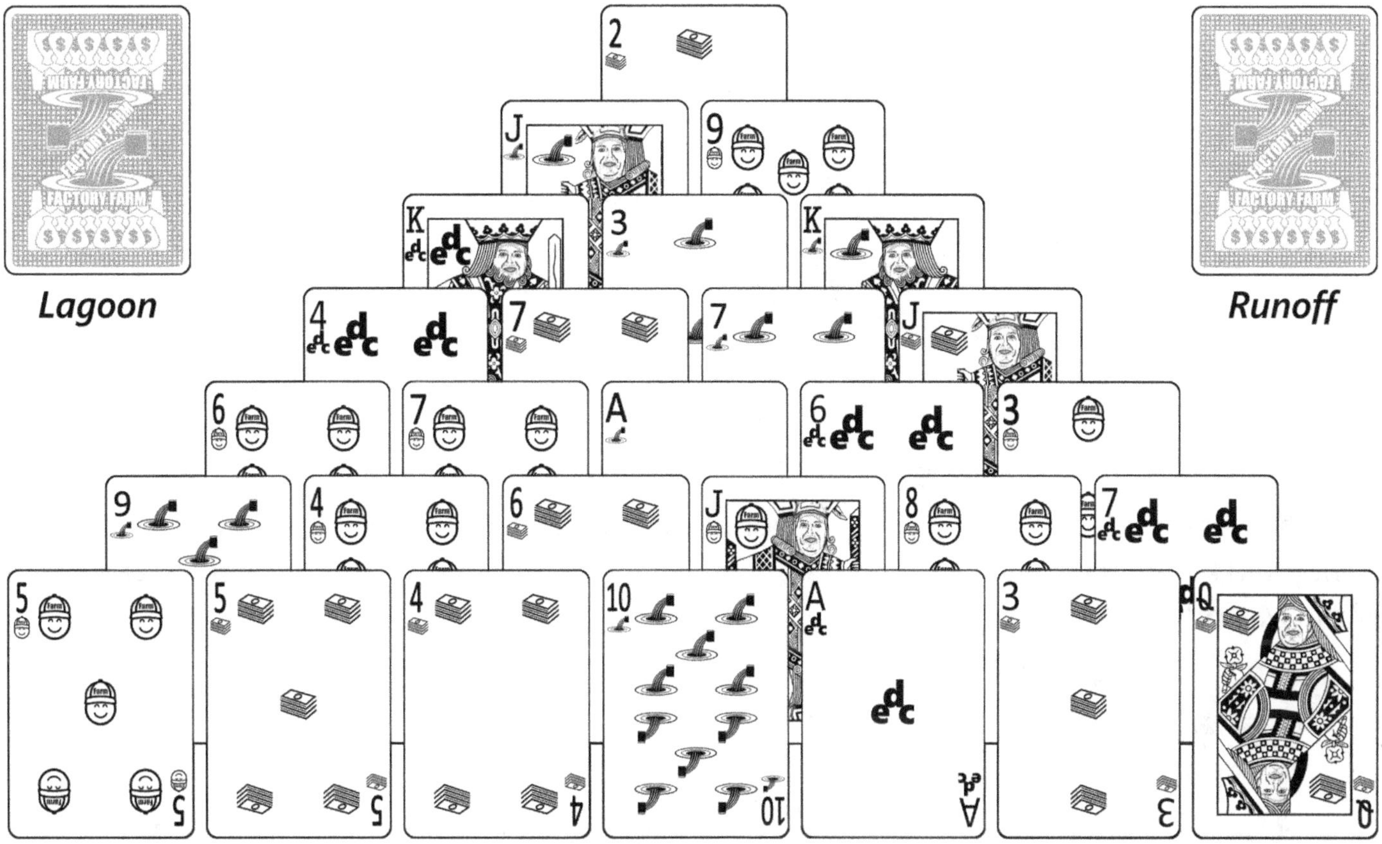
Lagoon
Runoff

The object of the game is to remove all the cards from the *lagoon*, *nutrient plan*, and *runoff*. Cards are removed singly or in pairs that add up to 13, regardless of suit. Kings count as 13, Queens as 12, Jacks as 11, and all other cards as their face value (Ace = 1). These are the valid moves: K, Q+A, J+2, 10+3, 9+4, 8+5, 7+6.

Only cards that are completely visible are available for play. The top card of the *runoff* pile, if any, is also available for play. If the top card of the *lagoo*n cannot be played, move it to the *runoff* pile.

When all the *lagoon* cards have been played or moved to the *runoff* pile, you can move all the cards from the *runoff* back into the *lagoon*. In this way you can go through the deck three times.

The game is won if you remove all the cards from the *nutrient plan*, *lagoon*, and *runoff*. But remember: Solitaire is like a Nutrient Plan — you don't have to follow the rules when nobody's watching.

MISFORTUNE TELLER

"It's your fault for living there."

INSTRUCTIONS:

1. Copy the printed square on the following page and cut out on the dotted lines.

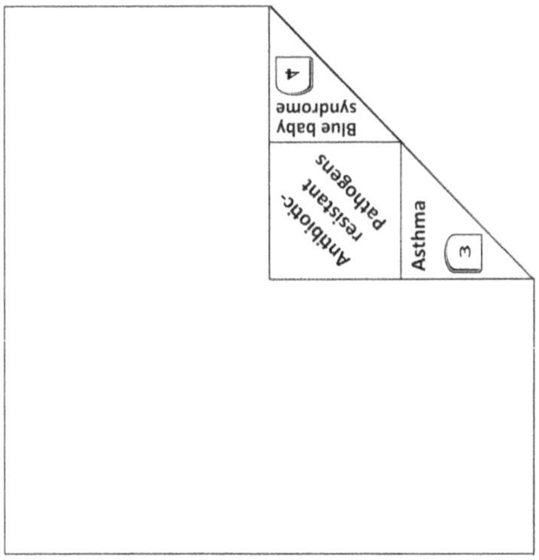
Blue baby syndrome
4
Antibiotic-resistant pathogens
Asthma
3

2. With the text side facing down, fold all four corners of the Misfortune Teller.

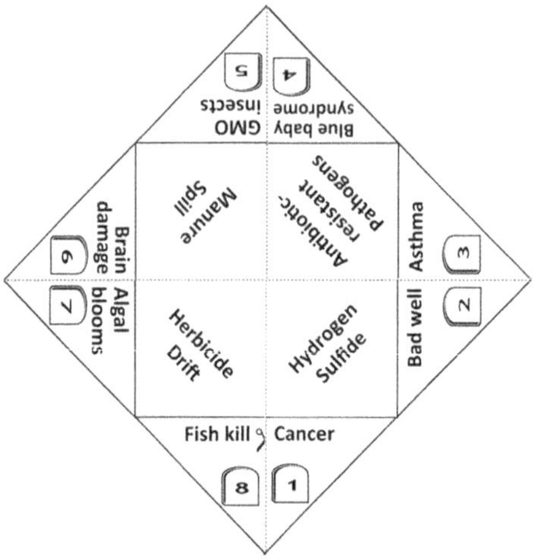
Blue baby syndrome
GMO insects
4
5
Manure Spill
Antibiotic-resistant pathogens
Brain damage
6
Asthma
3
Algal blooms
7
Herbicide Drift
Hydrogen Sulfide
Bad well
2
Fish kill
Cancer
8
1

3. It should look like this.

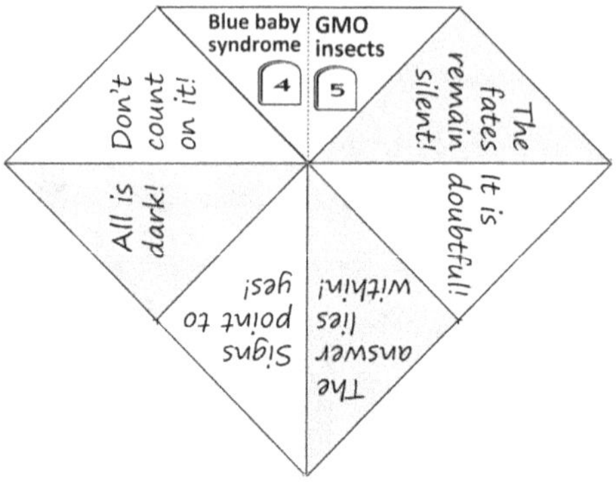
Blue baby syndrome
GMO insects
4
5
Don't count on it!
The fates remain silent!
All is dark!
It is doubtful!
Signs point to yes!
The answer lies within!

4. Flip paper over and fold all four corners again.

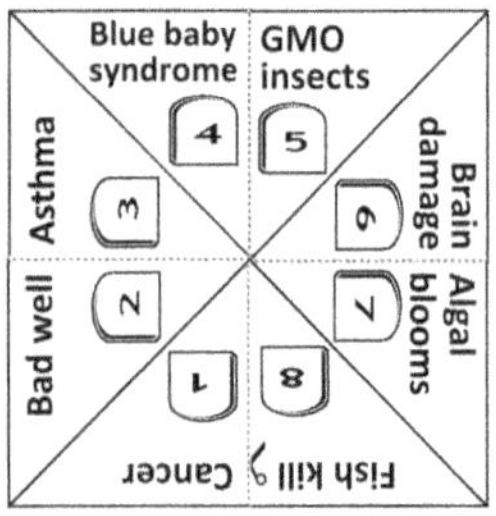
Blue baby syndrome
GMO insects
4
5
Asthma
3
Brain damage
6
Bad well
2
Algal blooms
7
1
8
Cancer
Fish kill

5. It should now look like this.

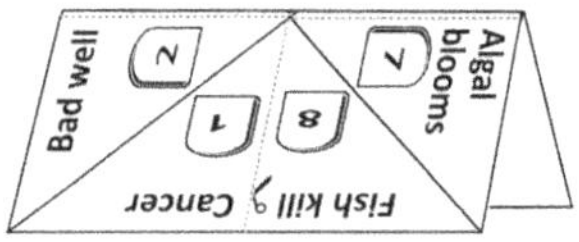
Bad well
2
1
8
7
Algal blooms
Cancer
Fish kill

6. Fold in half as shown.

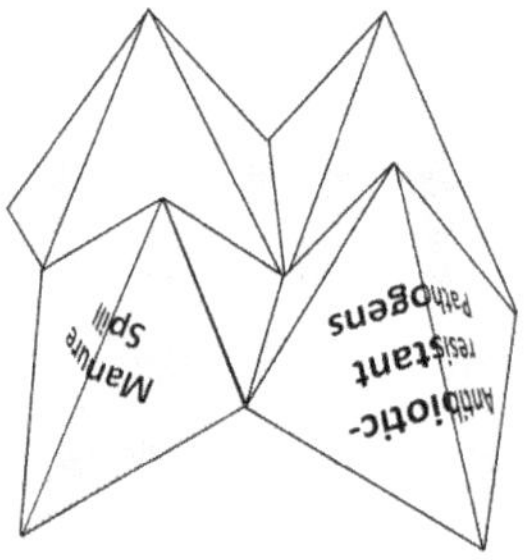
Manure Spill
Antibiotic-resistant Pathogens

7. Place your fingers under the four paper flaps and work back and forth to form creases.

HOW TO PLAY:

Begin with the thumb and index fingers of each hand in the four pockets of the Misfortune Teller.

Have the person whose fortune is being read pick one of the factory farm ills on the top four flaps.

Spell out the letters of the first word by alternating a pinching and pulling motion with the Teller.

Q: How do you kill antibiotic-resistant pathogens? A: ***We're working on it!***

Each pinch will expose four of the numbers on the inner flaps, and each pull will expose the other four numbers.

After spelling out the word, the Teller will be showing one of the sets of four numbers. The other player will then pick one of those numbers.

Pinch and pull the Teller while counting out that number. Once the number has been counted, four numbers will be exposed.

After one is picked, the fortune under that number is read.

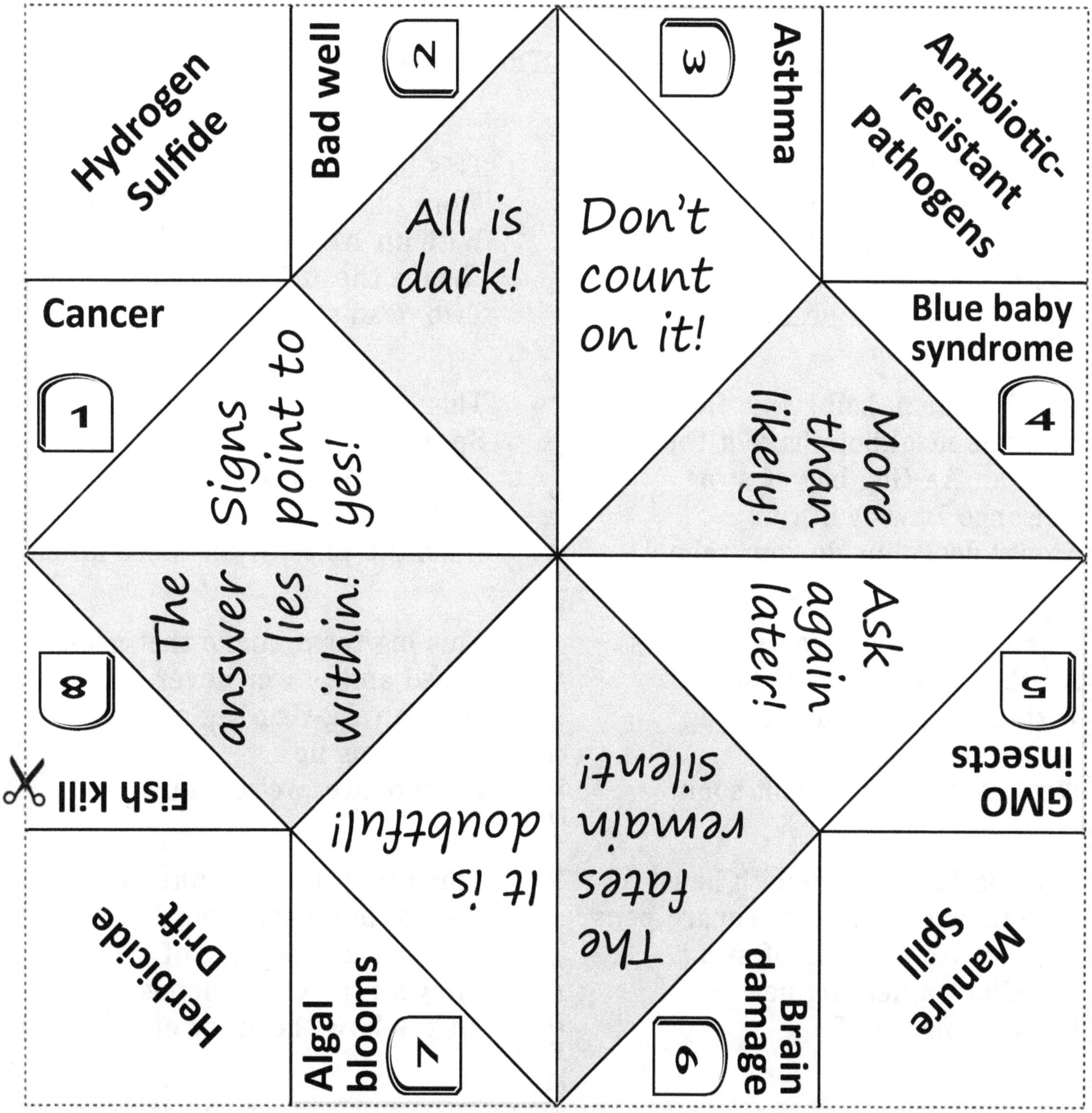

Hydrogen Sulfide
Bad well
Asthma
Antibiotic-resistant Pathogens
Cancer
Blue baby syndrome
GMO insects
Fish kill
Herbicide Drift
Algal blooms
Brain damage
Manure Spill
1
2
3
4
5
6
7
8
All is dark!
Don't count on it!
Signs point to yes!
More than likely!
The answer lies within!
Ask again later!
It is doubtful!
The fates remain silent!

Q: What do you do if you see the herbicide sprayer man? A: Run away as fast as you can, man.

Pandemic Sing-a-long

From a rigorous scientific viewpoint, the only benefit industrial farming holds for global hunger is an antibiotic-reistant pandemic that reduces the world's population.

Hold your own Pandemic Sing-a-long while there are still people left to invite.

To the tune of "This Old Man"

This big farm, has a plan
Grabbing everything it can
With an Ag-Gag bag of swag
Factory Farming wins
This is how it all begins

This big farm, had swine flu
Now the neighbors have it too
With an Ag-Gag bag of swag
Nuisance Laws will rule
Empty desks inside your school

This big farm, stuffed with cows
All the drugs the law allows
With an Ag-Gag bag of swag
Resistome lagoon
We'll all be in deep shit soon

This big farm, filled with hens
Live their lives in foot-square pens
With an Ag-Gag bag of swag
Breeding something new
Reassorting HP flu

PANDEMIC • PANDEMIC • PANDEMIC • PANDEMIC

This big farm, it has clout
More than all the poor without
With an Ag-Gag bag of swag
"Leave those farms alone!"
Two dead sisters in your home

This big farm, they don't care
Spreading virus in the air
With an Ag-Gag bag of swag
Body bags are filled
Burn the piles of all those killed

This big farm, made that germ
Greed and power never learn
With an Ag-Gag bag of swag
Politicians lie
It's too late, we're gonna die

This big farm, they did well
Profits in a year from Hell
With an Ag-Gag bag of swag
Flowers for your friends
This is how the story ends

   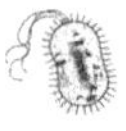

THE NEW FOOD PYRAMID

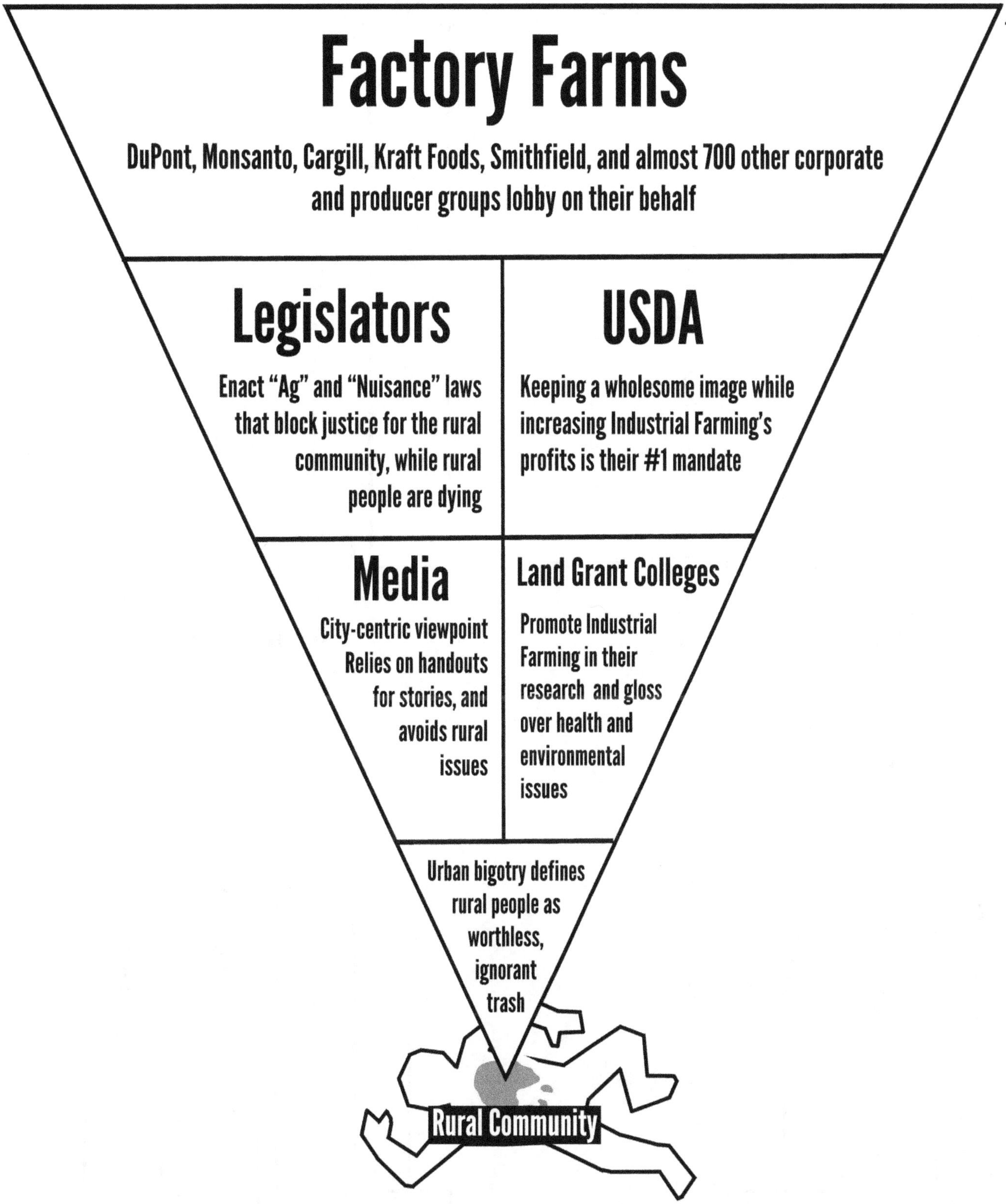
Factory Farms
DuPont, Monsanto, Cargill, Kraft Foods, Smithfield, and almost 700 other corporate and producer groups lobby on their behalf
Legislators
Enact "Ag" and "Nuisance" laws that block justice for the rural community, while rural people are dying
USDA
Keeping a wholesome image while increasing Industrial Farming's profits is their #1 mandate
Media
City-centric viewpoint Relies on handouts for stories, and avoids rural issues
Land Grant Colleges
Promote Industrial Farming in their research and gloss over health and environmental issues
Urban bigotry defines rural people as worthless, ignorant trash
Rural Community

Q: What are states doing about the liability of factory farms for injuring their neighbors?
A: Exempting them.

Color Me H-A-B

Color the lake phosphorus loading charts below by using a different color for each land use, then answer the question.

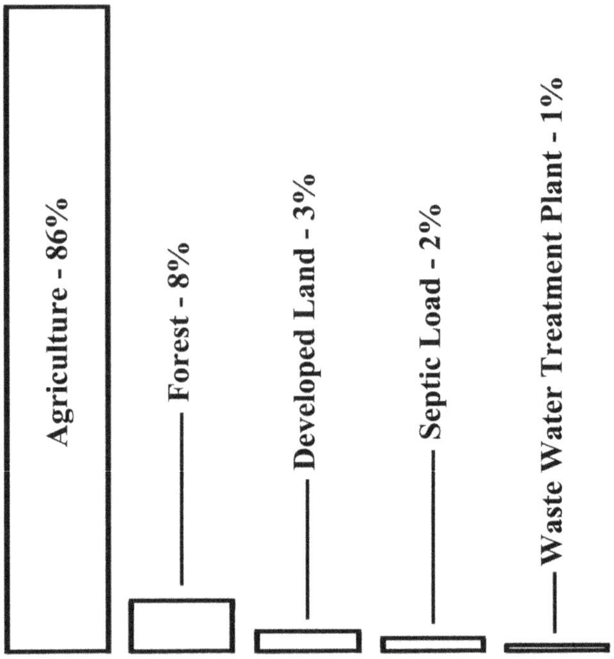
Agriculture - 86%
Forest - 8%
Developed Land - 3%
Septic Load - 2%
Waste Water Treatment Plant - 1%

Owasco Lake
Harmful Algal Blooms

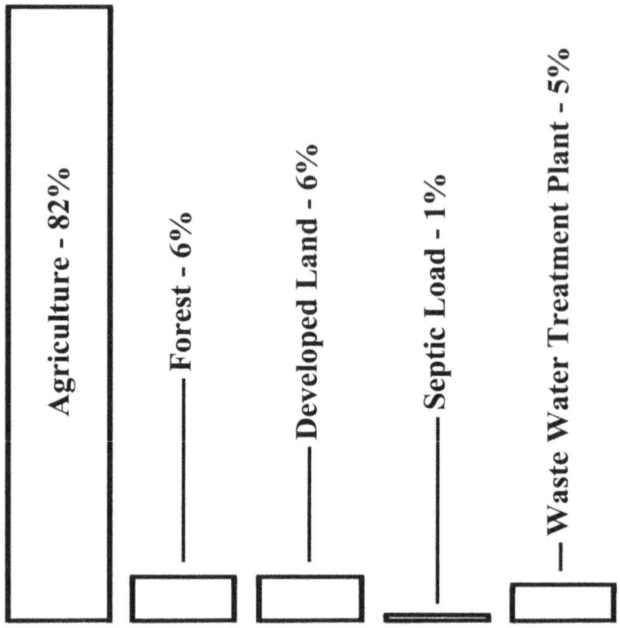
Agriculture - 82%
Forest - 6%
Developed Land - 6%
Septic Load - 1%
Waste Water Treatment Plant - 5%

Cayuga Lake
Harmful Algal Blooms

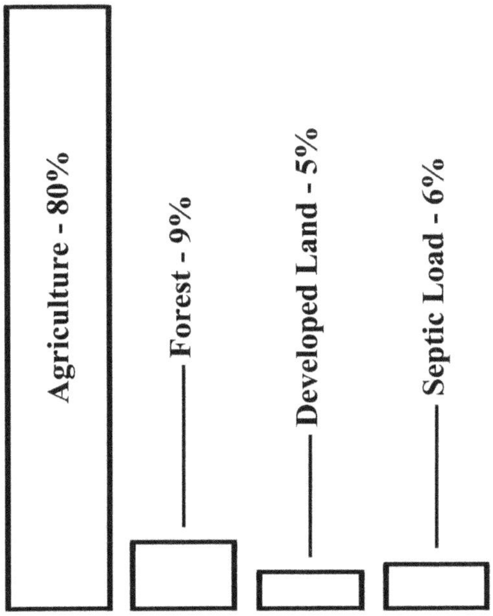
Agriculture - 80%
Forest - 9%
Developed Land - 5%
Septic Load - 6%

Skaneateles Lake
Harmful Algal Blooms

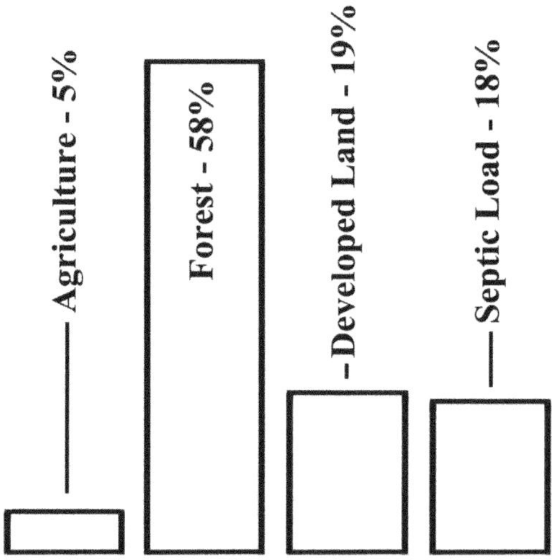
Agriculture - 5%
Forest - 58%
Developed Land - 19%
Septic Load - 18%

Lake George
No Harmful Algal Blooms

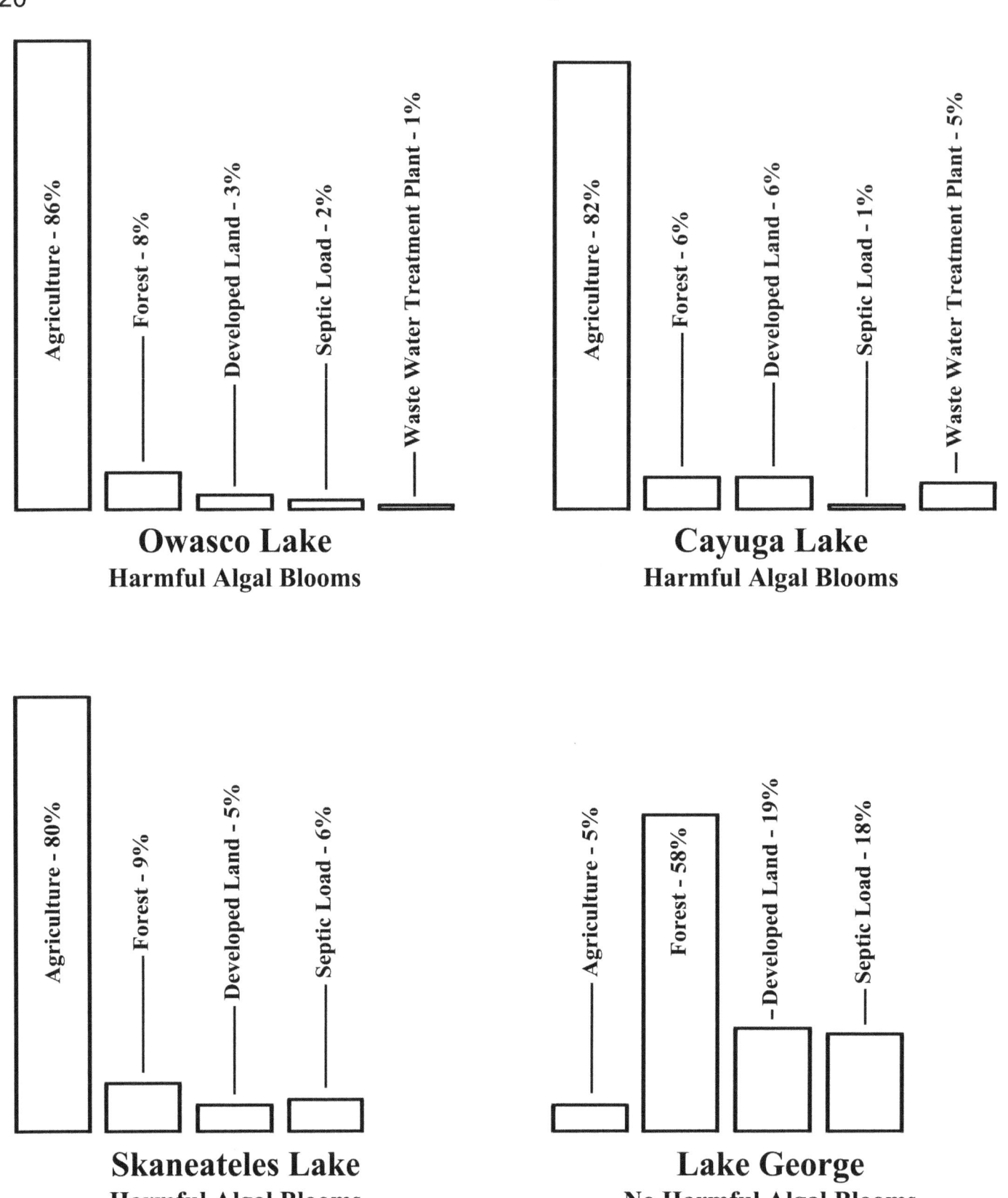

Q: Officials at the NYS Harmful Algal Bloom Summit could not find a pattern — can you?

Stopping by Woods on a Snow Melt Evening

Whose field this is I think I know.
He doesn't live around here, though;
He will not see me stopping here
To see manure covered snow.
Those urban folks must think it queer
And likely they will never hear
Of runoff in the woods and lake
That's happened several times this year.

They'll toss a quick dismissive shake
To show there must be some mistake.
Their city lives are much too chic
To deal with messes farmers make.
There was another spill last week,
But still the Governor won't speak,
And miles it flowed on down the creek,
And miles it flowed on down the creek.

(With apologies to Robert Frost, and none to industrial farming.)

FACTORY FARM DAYS

A celebration of modern farming practices

"From the killing-yards to your yard"

Today's industrial farms like to present themselves in a Mid-century pastiche of antique tractors, country-style pies, and a "Dairy Queen" and her court. It's time to peel back the packaging and show the tourists what rural life is really like.

The Fume Room

A kiddie-pool filled with real liquid manure is the key to this "fun house" event. Town residents get 15 seconds to experience a slice of modern rural life. There is no set time limit for rural residents — it's their fault for being there.

Note: If *Farm Days* are held within an Agricultural District, promoters, booth operators, and game vendors are protected from any liability by nuisance laws.

Herbicide "Duck-n-dodge"

A perennial favorite for kids of all ages. The object is to stay within the boundaries of the drift area while herbicide applicators try to "broad spectrum" the contestants. Prizes awarded to the survivors include tickets for the decontamination shower, N95/NIOSH Respirators, and a $100 gift certificate for the local urgent care facility.

Not recommended for the elderly, or those with walkers.

The Fly Dome

"More flies than you can shake a swatter at." Over half-a-million blow flies fill the inside of this spacious tent, whizzing and buzzing in an ever moving pattern of pestilence. Major elements of this educational family fun attraction include the "mound-of-maggots" offal pit, the hanging fly-paper obstacle course, and the science table, with its fly life-cycle display and forensic corpse photo board.

Spoiler: Keep your food and drinks covered. A cascade of charred flies falling from the bug zapper [nicknamed the "fly-fall"] is a constant source of amusement for onlookers, and an unforgettable lesson for the unwary.

For a complete list of *Factory Farm Days* venues and their corporate sponsors go to: *industrialfarming.org*

Q: How can you tell if they've been applying liquid manure? A: How can you ask?

EASY MAZE

Nitrate Pollution Maze

The Law when farms pollute your well:
"Drink bottled water, or just sell."

Fowl Play says — *"I don't give a toot, when farms pollute."*

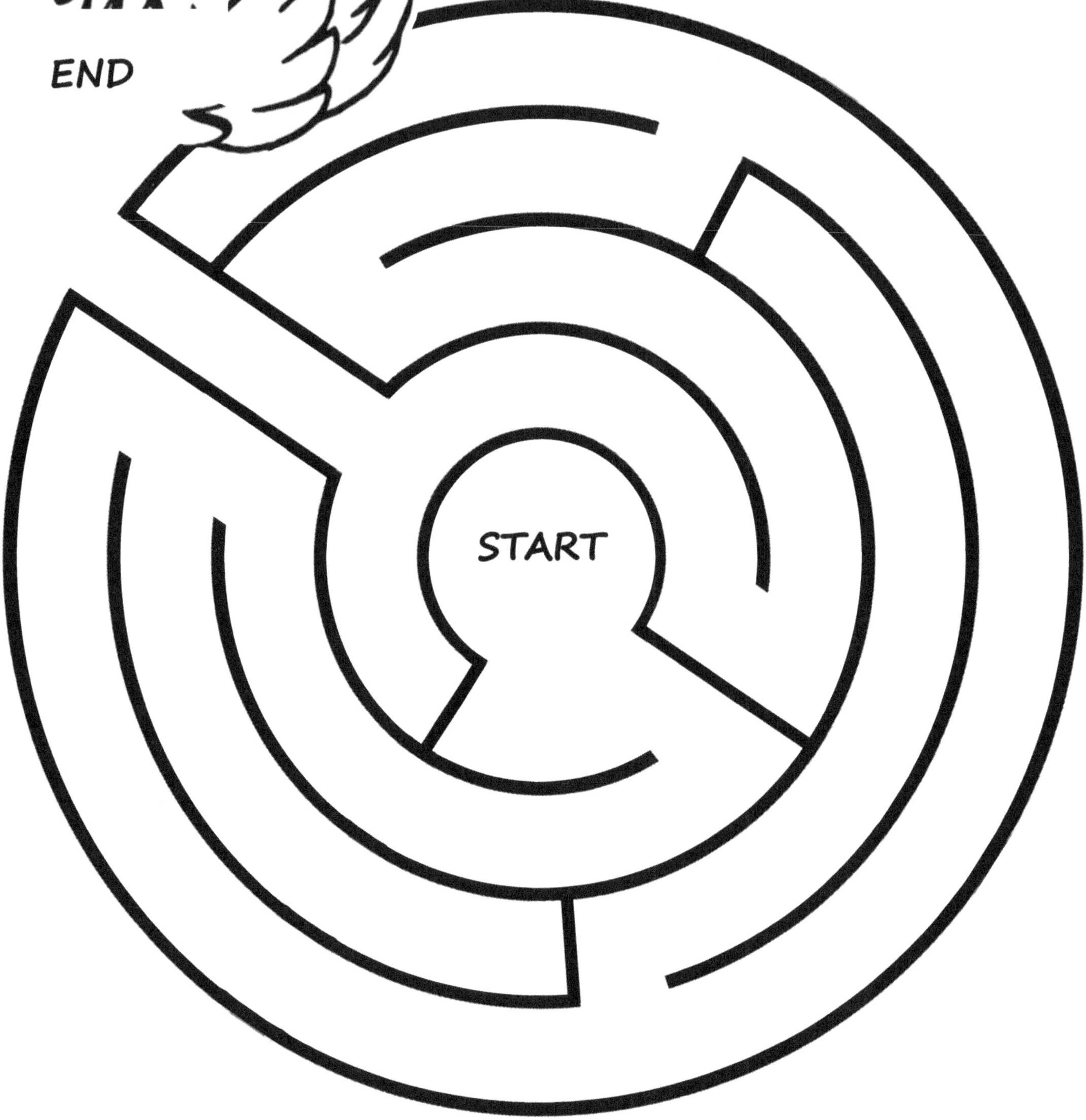
END
START

The Nitrate results were way over the top
Guide your family to safety, don't linger or stop!

CAUTION:
HARMFUL ALGAL BLOOM ALERT
Avoid contact with blooms.
Keep people and pets away from blooms.
Harmful algal blooms have been seen in this waterbody.
Blooms can make you and your pets sick.
DO NOT EAT FISH CAUGHT IN A BLOOM AREA
LEARN MORE www.countygov.loveourfarms.coverup.org
DANGER
DO NOT DRINK
THIS WATER

ALGAL BLOOM WARNING SIGNS

the urban dairy

Politicians are always touting the great economic benefits of agriculture, and yet the only "green" things around many of our cities are the lakes.

The *Urban Farm Initiative* is working to put Industrial Farming back in the industrial heart of our country.

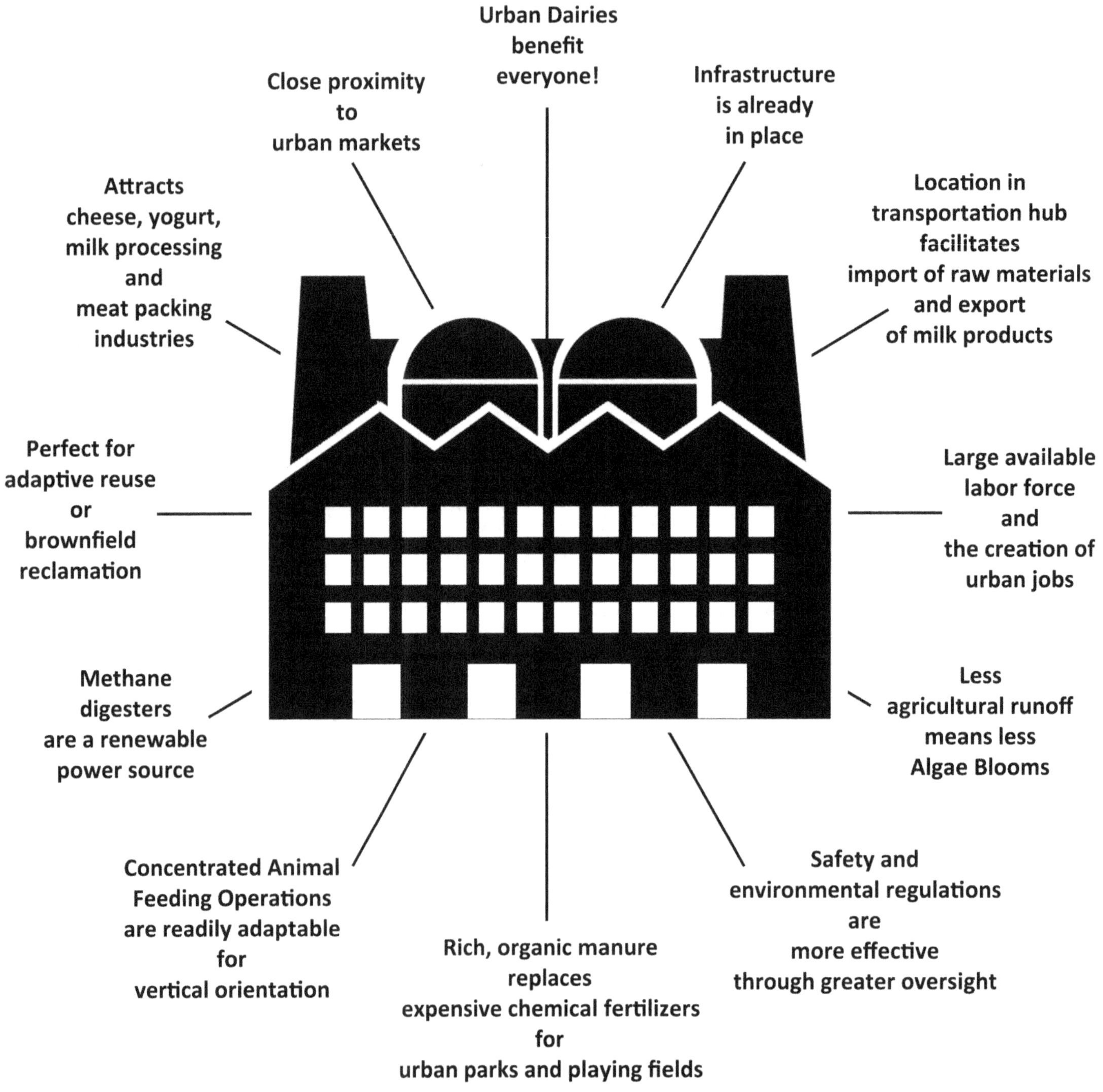
Urban Dairies benefit everyone!
Close proximity to urban markets
Infrastructure is already in place
Attracts cheese, yogurt, milk processing and meat packing industries
Location in transportation hub facilitates import of raw materials and export of milk products
Perfect for adaptive reuse or brownfield reclamation
Large available labor force and the creation of urban jobs
Methane digesters are a renewable power source
Less agricultural runoff means less Algae Blooms
Concentrated Animal Feeding Operations are readily adaptable for vertical orientation
Rich, organic manure replaces expensive chemical fertilizers for urban parks and playing fields
Safety and environmental regulations are more effective through greater oversight

"Industrial means Urban, and the Urban Dairy is the future of Industrial Agriculture."

Grow urban. Grow jobs. Grow green.

Pass the Buck

Make a parcel by wrapping up a prize in many layers of paper. Put a slip of paper with an industrial farming excuse, and a small amount of money, between each layer. Excuses may include:

"That hasn't been proven" "We're not sure what's causing it"

"It's a hundred-year flood plain" "There's no simple solution"

"A naturally occurring . . ." "They're important to the economy"

Sit the children in a circle, give one of them the parcel and when the music starts, get them to hand it around the circle.

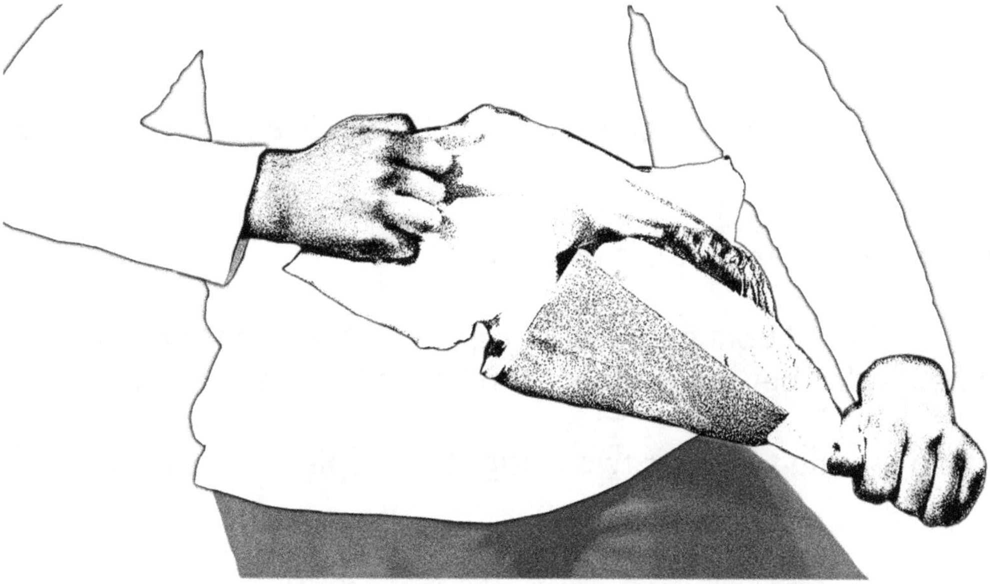

When the music stops, the child who's holding the parcel opens the first layer, reads the excuse, and collects the money.

Continue until one child unwraps the prize.

Farm-cheesi
The "Game of Flies"

Farm-cheesi is usually played with a pair of dice and the goal of the game is to move each of one's maggots home to the center space. The game board should be positioned so that the player's "carcass" [numbered circle] is to their right. Each player places the four maggots with that same number on their carcass.

Maggots enter play onto the darkened safe space to the left of their carcass and continue their life cycle counter-clockwise around the board, becoming pupae home path leading to their egg-laying adult blow fly in the center.

Each player rolls a die [spinners can be used instead of dice — see *Escape the Factory Farm* for their construction] and the highest roller goes first. Subsequent play continues to the player on the left. On each turn, players throw both dice and use the numbers shown to move their maggots around the board. If an amount on one or both dice cannot be moved, that amount is forfeited.

The player must use as much of the dice total as possible. If the player cannot use both numbers, they must use the highest of the numbers they can use.

Entering maggots:

A player may enter a maggot only by throwing a five or a total sum of five on the dice. Each time a five is tossed, the player must start another maggot if available.

Capturing:

Any maggot that is not on a safe space or part of a blockade can be captured by an opposing maggot.

(1) The captured maggot is sent back to its carcass.

(2) The player is awarded 20 bonus spaces for capturing the opposing maggot. The 20 spaces may not be divided between maggots and must be moved if it is possible.

Team Rules: If the opposing team has two maggots on a player's exit area, the player cannot exit.

Blockades:

When two maggots occupy the same space, they prevent any maggots behind the two from advancing past the blockade. This includes blocking any maggots from leaving their carcass. The two pieces that form the blockade may not be moved forward together to form a new blockade on the same roll.

No more than two maggots can occupy any one space. Two maggots of different numbers never occupy the same space except at the moment one maggot captures another.

Safe spaces [Safety Spaces]:

The dark spaces are safe spaces. A maggot may not be captured as long as it sits on one of these spaces.

The only exception is if a piece sits on the safe space where another player enters the board from their carcass. Those spaces are safe from all other players, but the maggot can be taken if the player whose carcass it is has a maggot on his carcass and rolls a 5 (as long

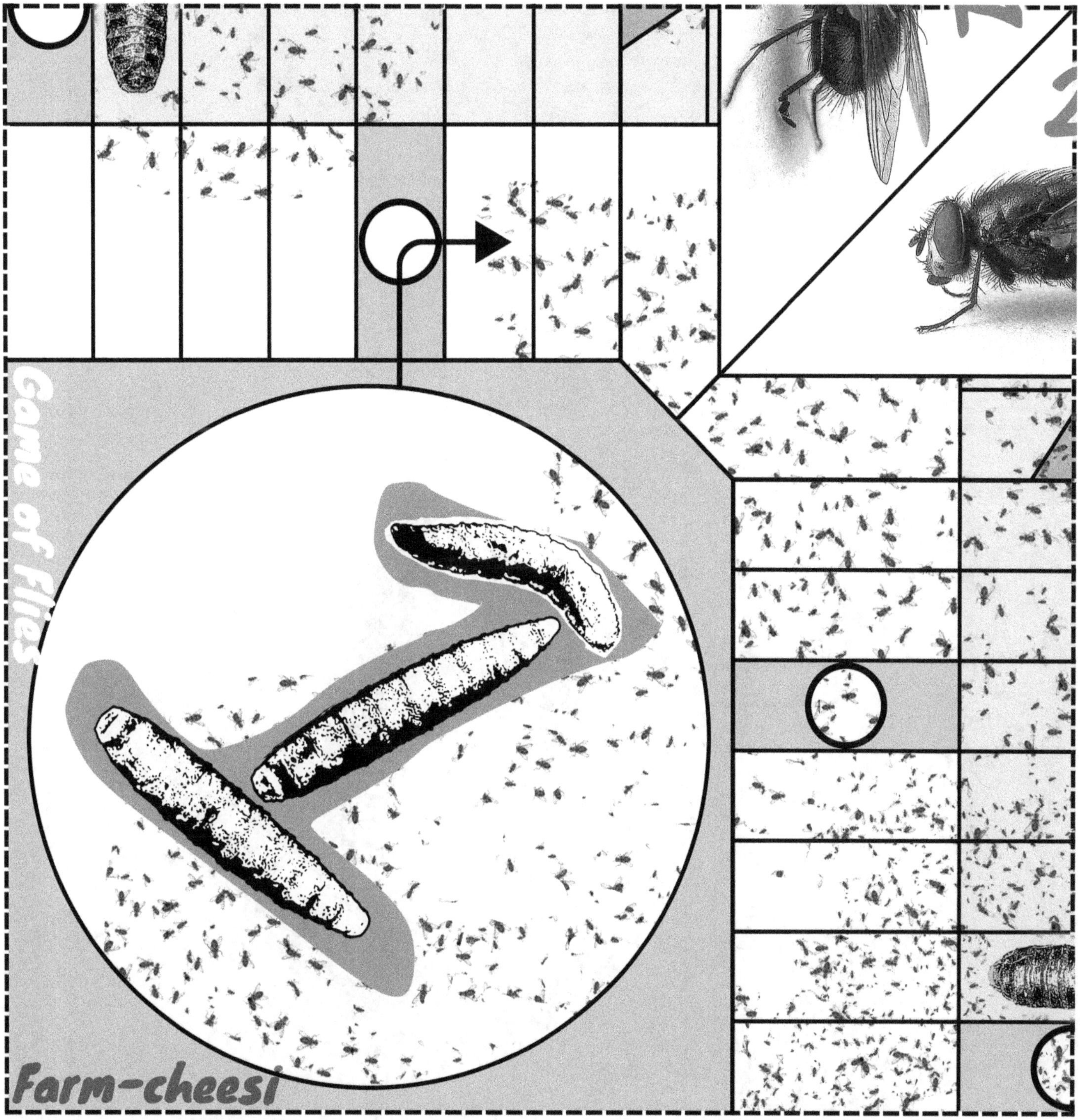
Game of Flies
Farm-cheesi

Farm-cheesi continued

as it isn't a blockade). For example: If you have a maggot sitting on another player's entry space and they roll a five and a maggot exits the carcass — they would capture your maggot and gain a 20 space capture bonus.

Note that two maggots of different numbers can never share a safe space. You can pass a single maggot on a safe space, but you cannot land on it, even temporarily, as part of your turn.

Two maggots that form a blockade are also safe.

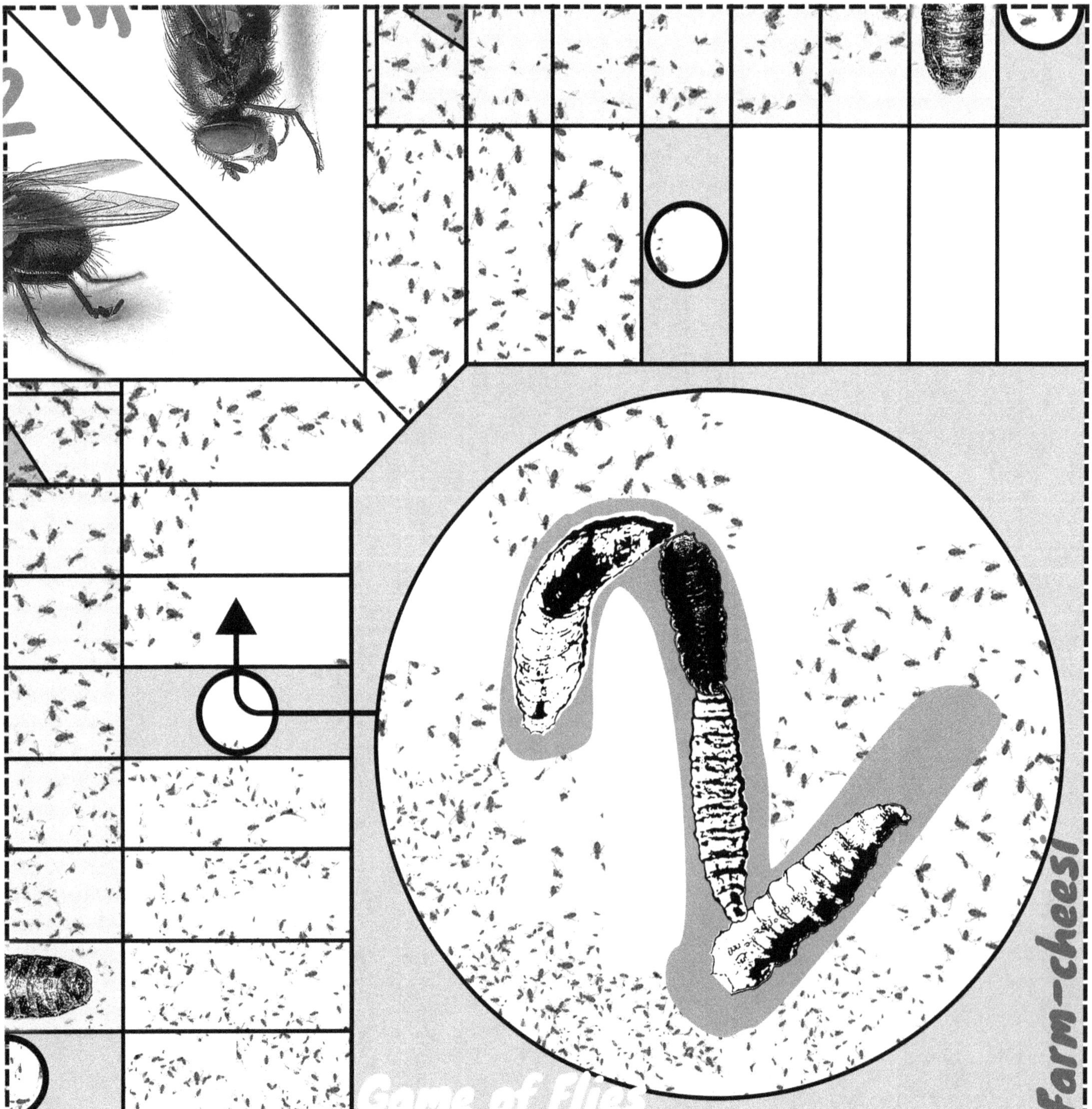
Game of Flies
Farm-cheesi

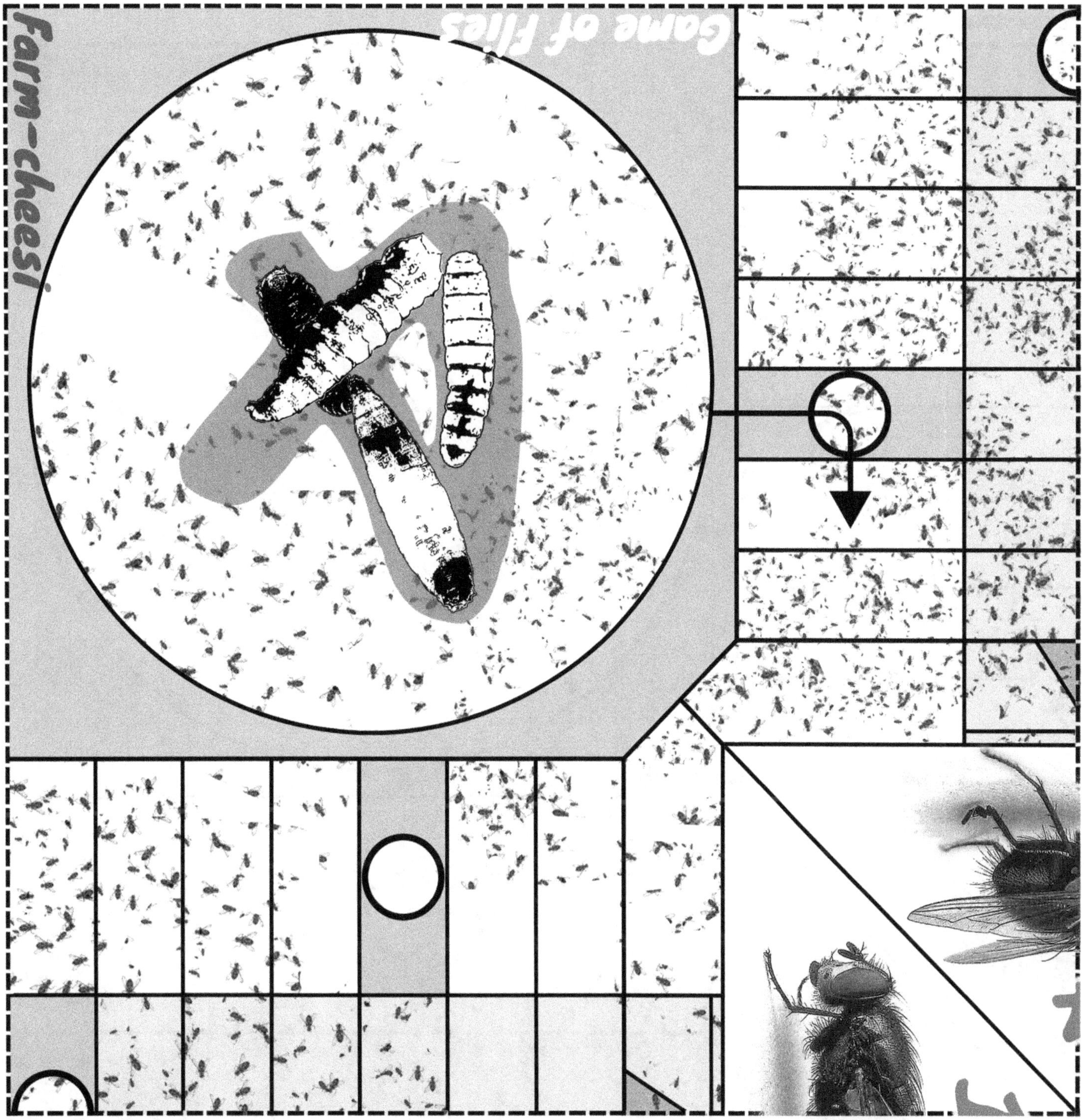
Farm-cheesi
Game of Flies

Doublets [doubles of the same number]:

When a doublet is tossed, the player gains another roll of the dice.

If all that player's maggots are off the carcass, the values on the reverse side of the dice [or opposite sides of the the spinners] are also used. For example: A player who rolls 6-6 can also move 1-1 in any combination. Therefore, when a doublet is tossed, the player has a total of fourteen spaces to move one or more maggots.

When all maggots are off the carcass, if a player rolls a doublet and cannot move all fourteen spaces, the player cannot move any spaces, and they must roll again.

Farm-cheesi continued

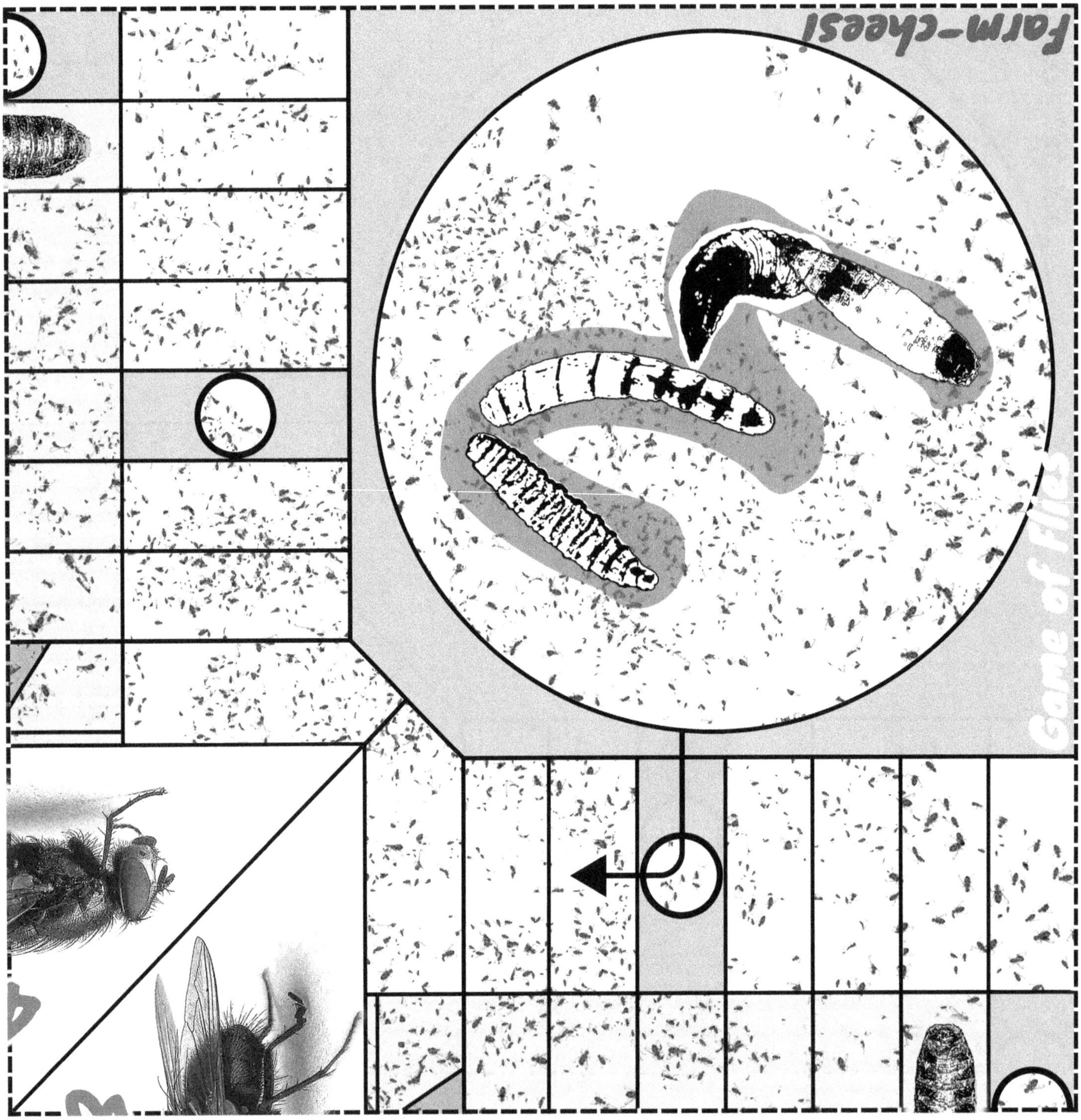
Farm-cheesi
Game of Flies

The third consecutive doublet rolled in one turn is a penalty. The player must move their maggot that is closest to home back to their carcass, and their turn ends.

A player cannot split doublets in order to enter home. A player can only enter home by rolling doublets if he is exactly that total number of spaces [always fourteen] from home.

Home:

Each player has their own home path and may not enter another's. When a maggot is on its home path, it can no longer be captured.

The center home space can only be entered by an exact throw of the die or dice. Home counts as a space.

When a maggot enter the center space by an exact count, that player is awarded ten movement points that can be used by any one maggot still in play at the end of their turn. If the bonus movement points cannot be used, they are forfeit.

Winning the game:

The first player to get all four maggots home wins and must yell ***"Farm-cheesi!"***

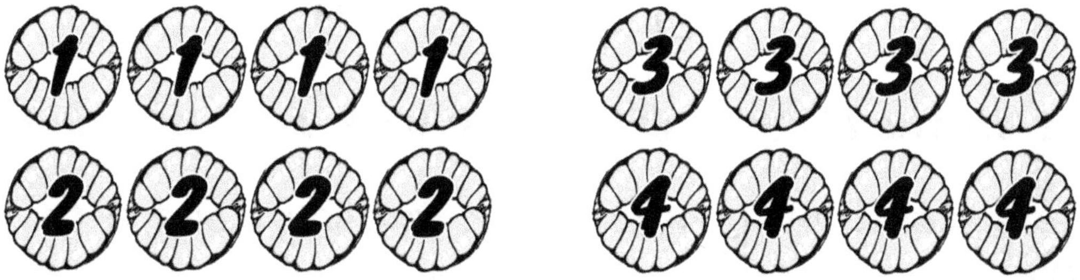
1
1
1
1
3
3
3
3
2
2
2
2
4
4
4
4

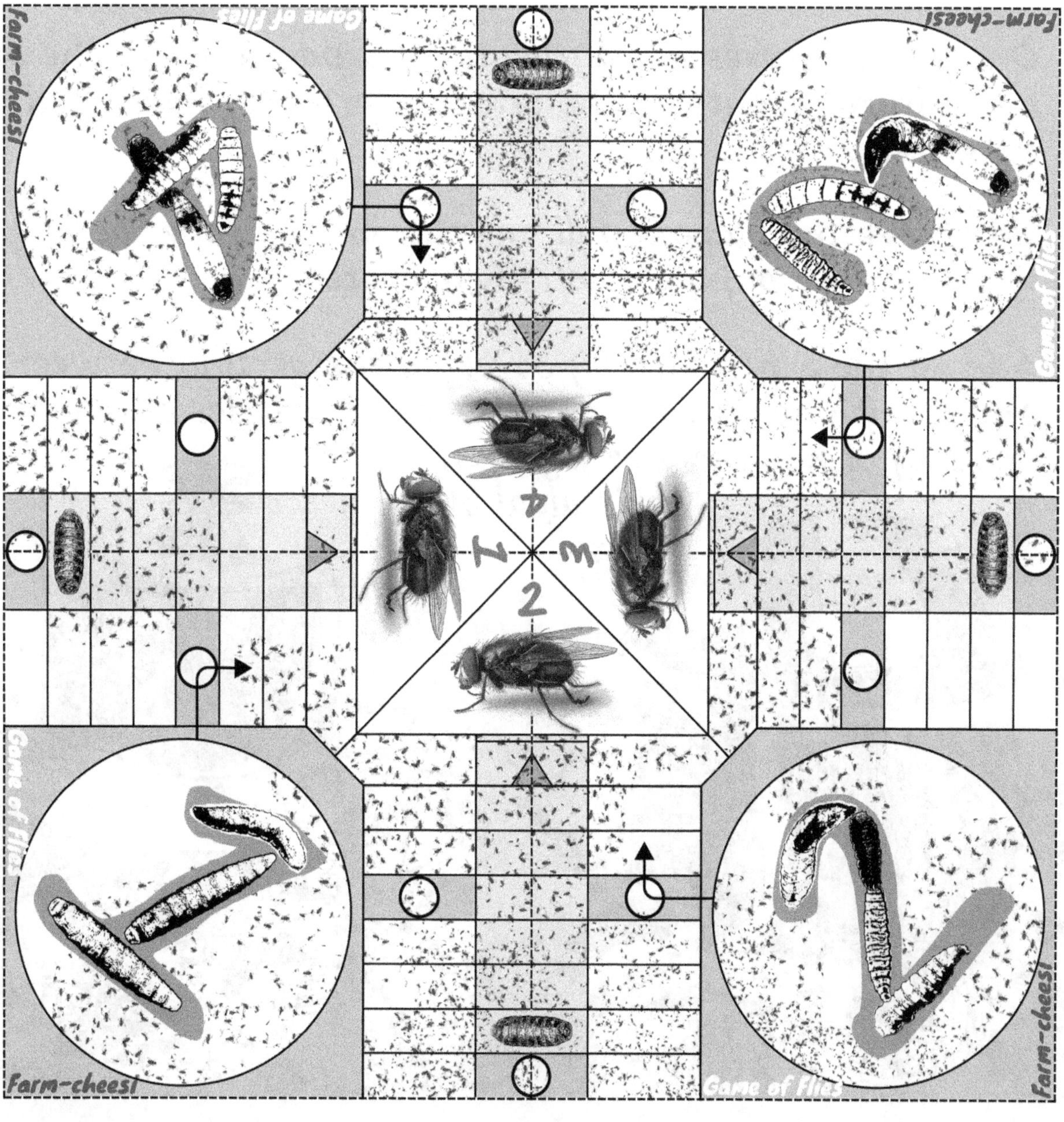
Farm-cheesi!
Game of Flies

Copy and cut out the maggot tokens for each player — tape the tokens to small coins for added stability.
Copy and cut out the four board pieces on the previous pages and assemble the board as shown.

Ecosystem Tug-of-War

Put a twist on the classic tug-of-war with a bit of role-playing:

1. Players stand on upended milk crates placed 6 to 10 feet apart.

   One player represents agricultural expansion and the other agricultural profits — the crates represent a balanced ecosystem.

2. Then have the players pull or relax on the rope, attempting to force their opponent off their crate.

   Add to the experience by covering the ground between the crates with newspaper and magazine articles on hurricane flooding of hog manure lagoons, toxic algal blooms, etc.

Hog Farm Sing-a-long

To the tune of "I've Been Working on the Railroad"

I've been living near a hog farm
All the livelong day
I've a swatter that's a yard wide
Just to brush the flies away
Can't you smell the shit their spraying
Speckling my siding and my lawn
And lagoons they're underplaying
Overflow in every storm
Hog farm won't you go
Hog farm won't you go
Hog farm won't you go awa-a-a-ay
Hog farm won't you go
Hog farm won't you go
Hog farm won't you go away

Someone's in the bed with Congress
Someone's in the bed I know oh-oh-oh
Someone's in the bed with Congress
Handing out election dough, and saying
You can pocket this dough
You can pocket this dough oh-oh-oh
You can pocket this dough
Nobody will ever know
Don't you hear the whistleblowers
Rise up to fight for those unborn
Can't you hear the hog farms laughing
Spilling out their scorn

Someone's in the bed with Congress
Someone's in the bed I know oh-oh-oh
Someone's in the bed with Congress
Handing out election dough
You can pocket this dough
You can pocket this dough oh-oh-oh
You can pocket this dough
Nobody will ever know
Handing out election dough
Handing out election do-o-ugh

Book 'em

Biblio-tech

Sometimes, you *can* tell a book by its cover.

VOLUNTARY AGRICULTURAL GUIDELINES by Seymour Profits

MANURE HANDLING PRACTICES by Hugh Spillit

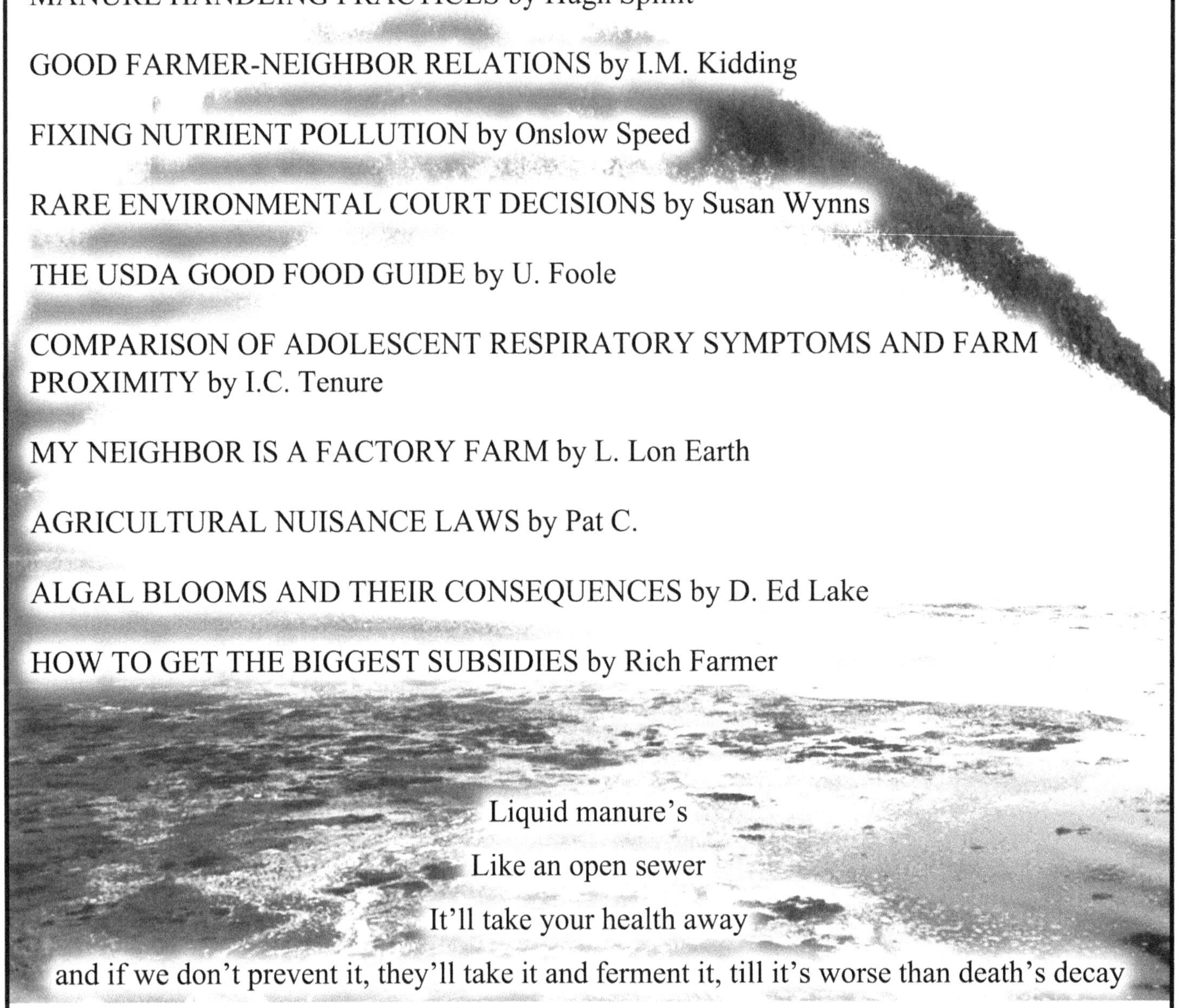

GOOD FARMER-NEIGHBOR RELATIONS by I.M. Kidding

FIXING NUTRIENT POLLUTION by Onslow Speed

RARE ENVIRONMENTAL COURT DECISIONS by Susan Wynns

THE USDA GOOD FOOD GUIDE by U. Foole

COMPARISON OF ADOLESCENT RESPIRATORY SYMPTOMS AND FARM PROXIMITY by I.C. Tenure

MY NEIGHBOR IS A FACTORY FARM by L. Lon Earth

AGRICULTURAL NUISANCE LAWS by Pat C.

ALGAL BLOOMS AND THEIR CONSEQUENCES by D. Ed Lake

HOW TO GET THE BIGGEST SUBSIDIES by Rich Farmer

Liquid manure's
Like an open sewer
It'll take your health away
and if we don't prevent it, they'll take it and ferment it, till it's worse than death's decay
Politicians won't stop it, it puts money in their pocket, and they never cared about us anyway
But if there's no solution to this factory farm pollution
Then the Silent Spring will settle here
To stay.

Fixing the Environment

A locked-door Maze

You are fighting your way through an unfamiliar maze of bureaucratic obstructions in an attempt to stop the building of a 27,000 hog factory farm in a flood plain just outside the town's conservation area. There is a maze of departments and officials, all carrying time penalty points. Can you make it to the hearing with the required documents within the 90 day limit? [Warning: There is no bureaucratic pathway to successfully block the farm's construction.]

■ Clerk on Vacation = 5 Days

▲ Expert Appraisal = 10 Days

) Technical Delay = 15 Days

★ Validate Petition = 20 Days

● Filling out Forms = 25 Days

♥ Room 203 = 30 Days

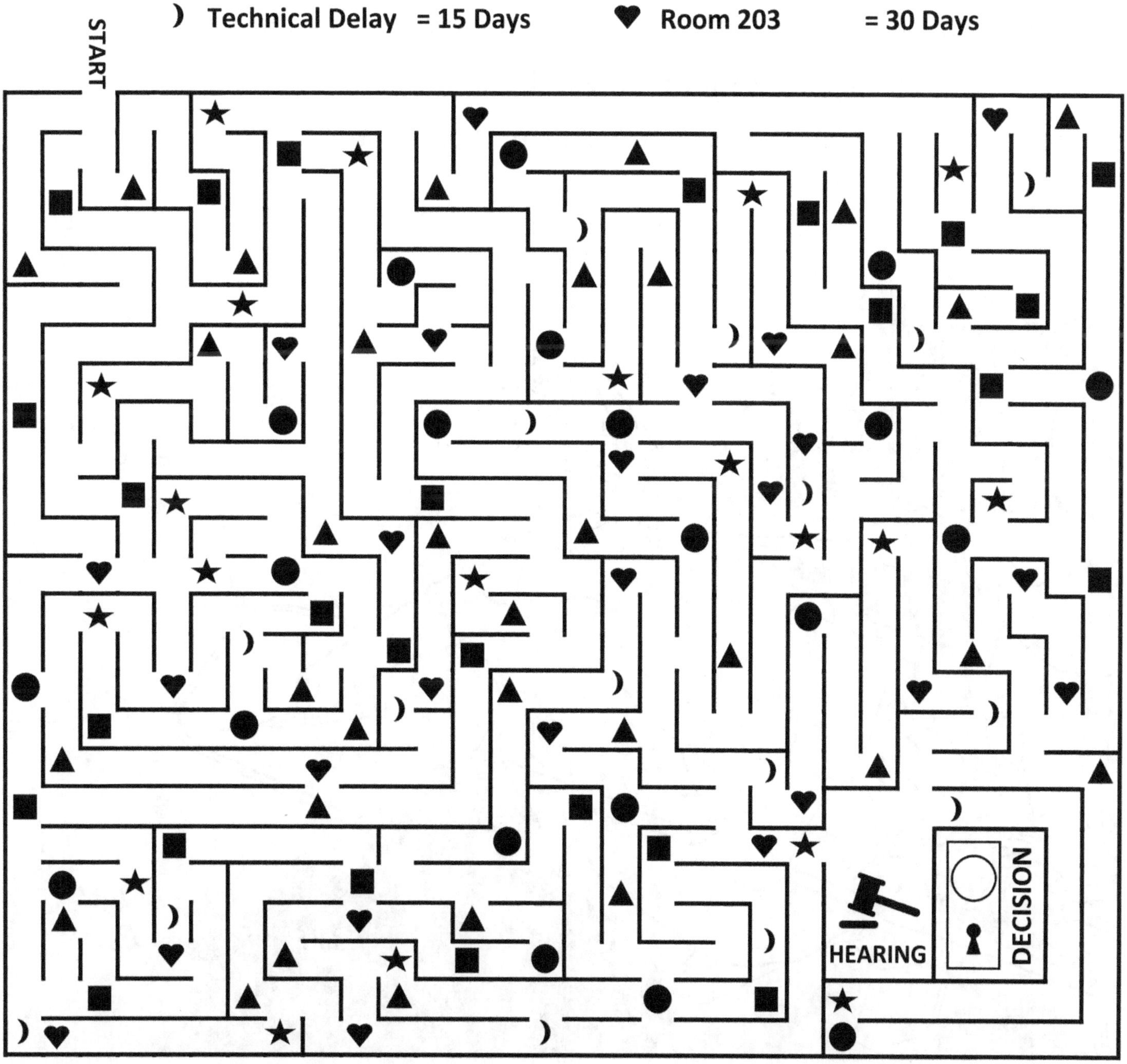
START
HEARING
DECISION

Fish Skimming

"Try it – If it's the last thing you ever do."

Fishing fun way beyond "green"

All you need is a pool skimmer and a pond or waterbody near a factory farm.
[Chemical resistant PVC or Nitrile gloves, boots, goggles, respirator and protective clothing is recommended.]

The Rules:

Objects must be completely removed from the pond and placed in the player's "treasure trove." At the end of the game, the player with the most points wins!

NO throwing rocks or farm debris into the pond to move objects closer.

NO pushing players into the pond.

NO unnecessary disturbance of blow flies.

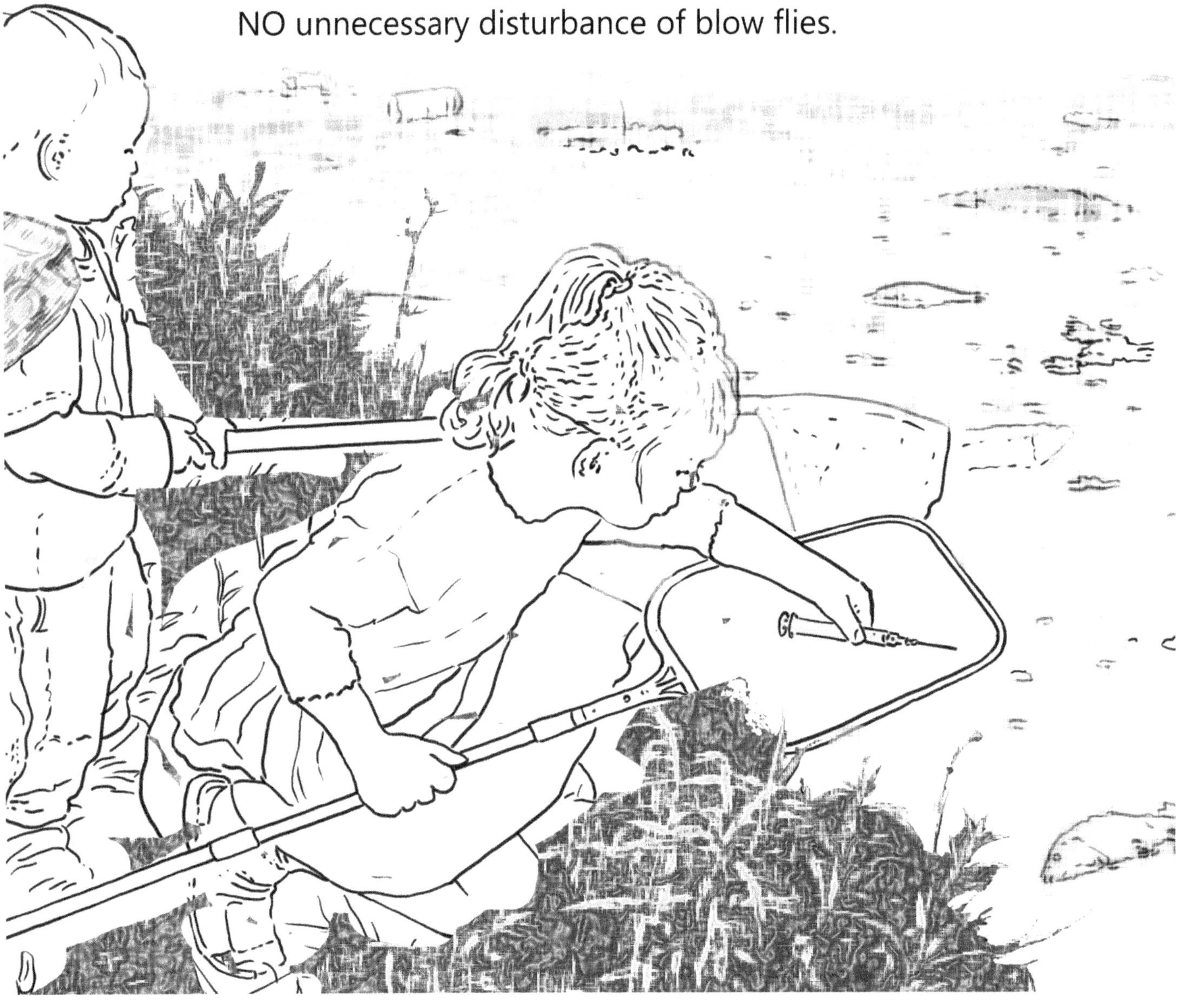

Scoring:

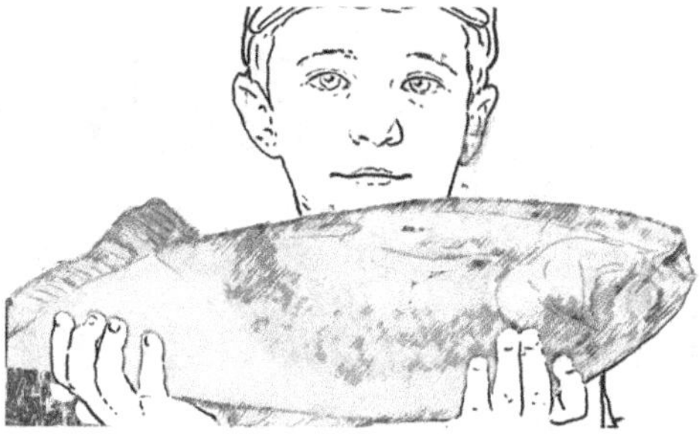

- Biggest fish - 150 points
- Most fish - 75 points

Bonus points for factory farm debris:

- Any antibiotic material - 10 points each [syringes - 25 points]
- Any agricultural enhancers - 8 points each [including hormones, GMO products, and any legible packaging]
- Plastic agricultural containers and bottles - 7 points each
- Corrugated cardboard - 5 points each box
- Visual sighting of discarded refrigerators, air conditioners, or any other used farm equipment, or any part thereof [must be partially submerged] - 5 points awarded to the first player to call out each object.

No points are given for wire, plastic twine, etc. as they may become entangled and disturbances could release toxic gases and particulates from the sediment.

[Safety tip: Dispose of or clean all gloves, boots, goggles, and skimmers after use. Bag clothes for washing and immediately bathe or shower. Keep all dogs and pets away from the "water," fish, and any used or collected articles.]

Variations:

Agal Skimming – this variant gaining in popularity due to the dramatic increase of algae in waterbodies receiving factory farm runoff.

CAUTION:
HARMFUL ALGAL BLOOM ALERT
Avoid contact with blooms.
Keep people and pets away from blooms.

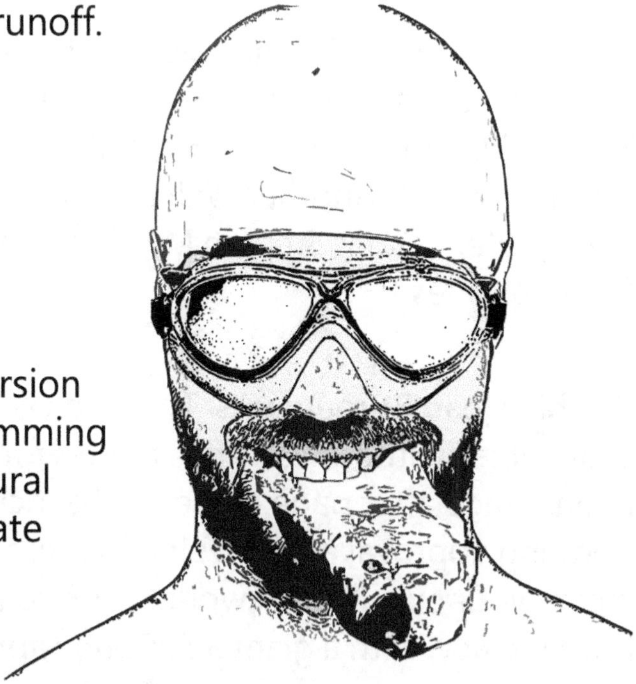

Extreme Fish Skimming - an over-the-top version of the children's game with competitors swimming unprotected and retrieving fish and agricultural debris with their teeth. Winners must celebrate outside the bar, and at a distance of at least 30 feet from any food preparation.
Hug at your own risk.

Factory Farm Bookshelf

"The best in industrial farming books, music, and games."

New and popular:

Make Mine Pink Slime An author's argument for acceptance of the new paradigm. "After years of being limited to 'old school' organic products, thanks to the USDA's leadership, consumers are now able to choose from a wide range of agricultural offerings, all with the trusted 'organic' label. Industrial farming has done an incredible job of expanding the organic market, and I'm proud they picked me to write this book!" — *AW*, Cooperative Extension Agricultural Writer

Ag Slang Flash Cards "Ag up" your water cooler chat with this informative flash card set. Did you know that "100 Year Storms" occur several times each year? Or that a "Family Farm" is a multi-national corporation where the majority of stock is owned by one family? Just follow the instructions, and in a matter of days you'll be understanding what the "rich people" mean when they say: "I'll get the hog farm expansion approval before the next *100 year storm*." and "If the governor can't push back those regulations for our *family farms*, he won't be in office come next election." Even if you're a party of the third part, liven up your party with authentic *Ag Slang Flash Cards*.

The Factory Farm Songbook Words and music from *Spillboard's* Top 100 Industrial Farming Songs of All Time — featuring: "Killing Me Softly With His Shit," "Spray-drops Keep Fallin' on My Head," "Hey Rube," "Imagine No More CAFOs," "As Slime Goes By," "Ruined River," "Another Well Bites The Dust," "I Want To Own Your Land," "You Blight Up My Life," "Don't Step on My Blue Baby's Shoes," and 90 more classic dystopian farming tunes.

Bobble Head Bureaucrats This book is the recognized authority on those cult collectible "authorities." Price guides for plastic, ceramic *and* stamped metal bobble heads. Page after page of die-cut "punch and paste" multi-cultural paper-pushers, nodding approval at all levels of government. Printed diagrams and instructions to create your own "paper-works." Add to this an annotated history of factory farm bureaucracy, with a giant *3,500 page* appendix of equivocations and runarounds, and you have a book that's as hard to pick up as it is to put down.

A+ for Asthma (a nurse's heartwarming story) Young Jason B. would come to the nurse's office every morning with severe asthma symptoms — even though he taped up the windows and door to his bedroom at night. See his struggle through a nurse's eyes, as she fights to help him escape an Ag Ghetto of fish kills, fumes, and factory farms. Share their triumph as he graduates with honors and wins a scholarship to a college in the state capitol, where factory farm "nutrients" are banned as "toxic waste."

Factory Farm Origami Book From Blow flies to Bureaucrats, factory farms represent an "invasive species" subsector that is overturning the ecological balance of rural America. Tradition meets perdition in this book, as a "harmful farm-full" of villains are transformed into Japanese art. "Money-driven policy making and corruption have always been with us, but the *Factory Farm Origami Book* provides some new folds on an old wrinkle," joked Dr. E. Steward, just hours before his tragic accident.

Where's the Inspector? Children under seven can have hours of fun trying to find the Inspector in a variety of factory farm settings — animal mortality dumps, slurry runs, medical waste piles, spoilage trailers, manure pits, the owner's suite, and more. Every page has a different setting, but only one has the Inspector . . .can you find her?

The Ag Almanac "What's the record for herbicide drift on a windy day?" and "How far can a toxic gas plume travel?" *The Ag Almanac* answers these and many other agricultural questions in a refreshingly straightforward way. The statistics section includes such items as a year-by-year government subsidy flow-meter, and a size comparison of the Dead Zone to various states. And the *Almanac's* long-term weather predictions are based, not on global warming trends, or climate change scenarios, but on a simple-to-understand mix of government influence and career longevity. This book is an indispensable source for measuring our lack of stewardship.

The Child's Water Testing Picture Book Kids: Your parents trust their well water because it's been pure and healthy for generations, but times have changed. Help keep your family safe with *The Child's Water Testing Picture Book.* Follow the simple instructions and diagrams to collect your well water samples, then bring them to a laboratory for testing. It's best to test weekly, or even daily during the Spill Season. Remember the Law: "Home owners with their own wells are solely responsible for the quality and safety of their water." So if you die, it's *your* fault. With frequent enough testing, it's comforting to know that your family will probably, in most cases, be OK.

ORIGAMI FLY

Nothing says "factory farm" like a blow fly, so here's how to make your own traditional origami fly. Color or mark the flies and use as tokens in the *Farm-cheesi* game or, for a touch of factory farm chic, add to a barrette or pin to your clothes.

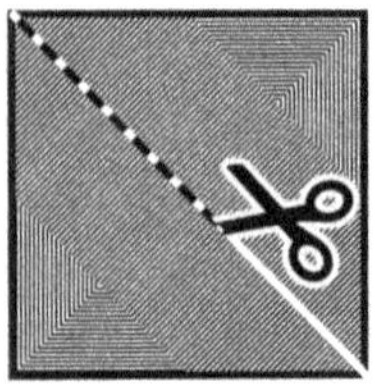

1. Copy and cut out the square design in the lower right hand corner of the page. Then cut it in half along the diagonal line to make two triangles. [Each triangle makes one fly.]

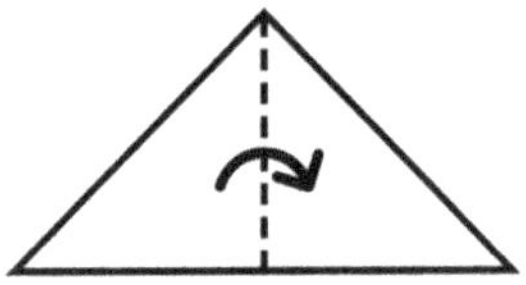

2. Turn one triangle to the plain side, fold in half and unfold.

3. Fold up the two bottom corners so that the corners meet at the peak of the triangle.

4. Fold the two corners back down at a bit of an angle so the corners are slightly away from one another, like the wings of a bug.

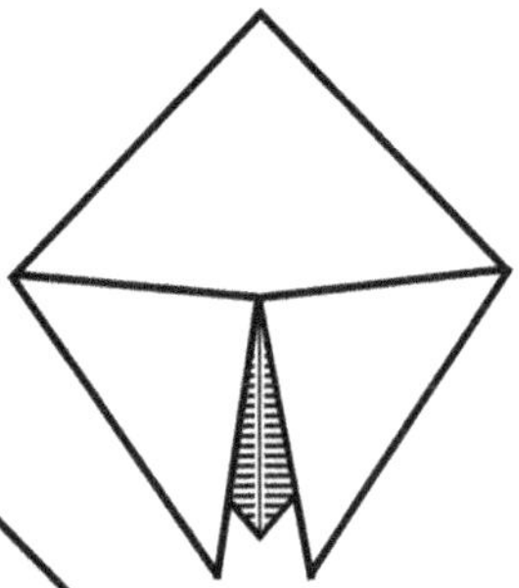

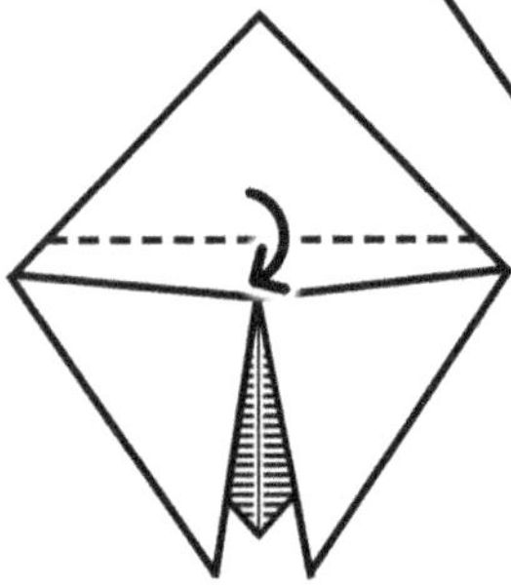

5. Fold down the top of the triangle; don't fold down exactly in half, leave a small gap as shown. Unfold.

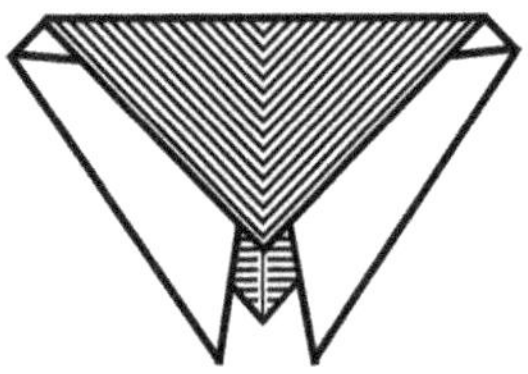

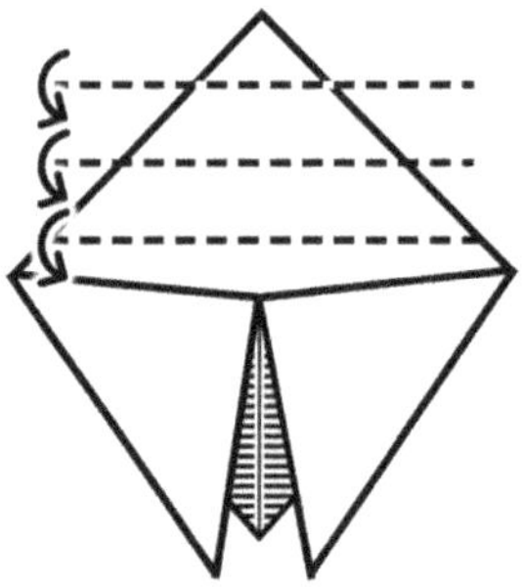

6. Fold down the top section of the triangle into thirds. Then fold down once more along the crease made in step 5.

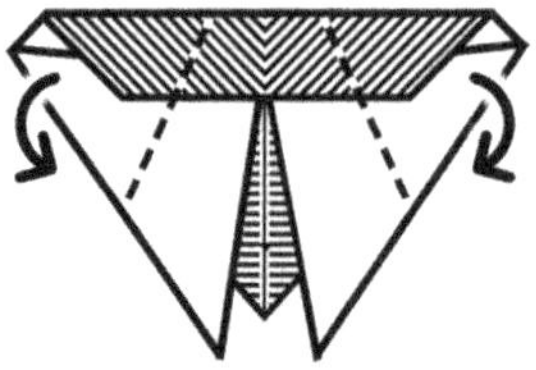

7. Fold back the left and right sides of the model to form the body of the fly.

8. Fold the tips of the wings a little so they look more like a fly's.

EASY MAZE

GMO Maze

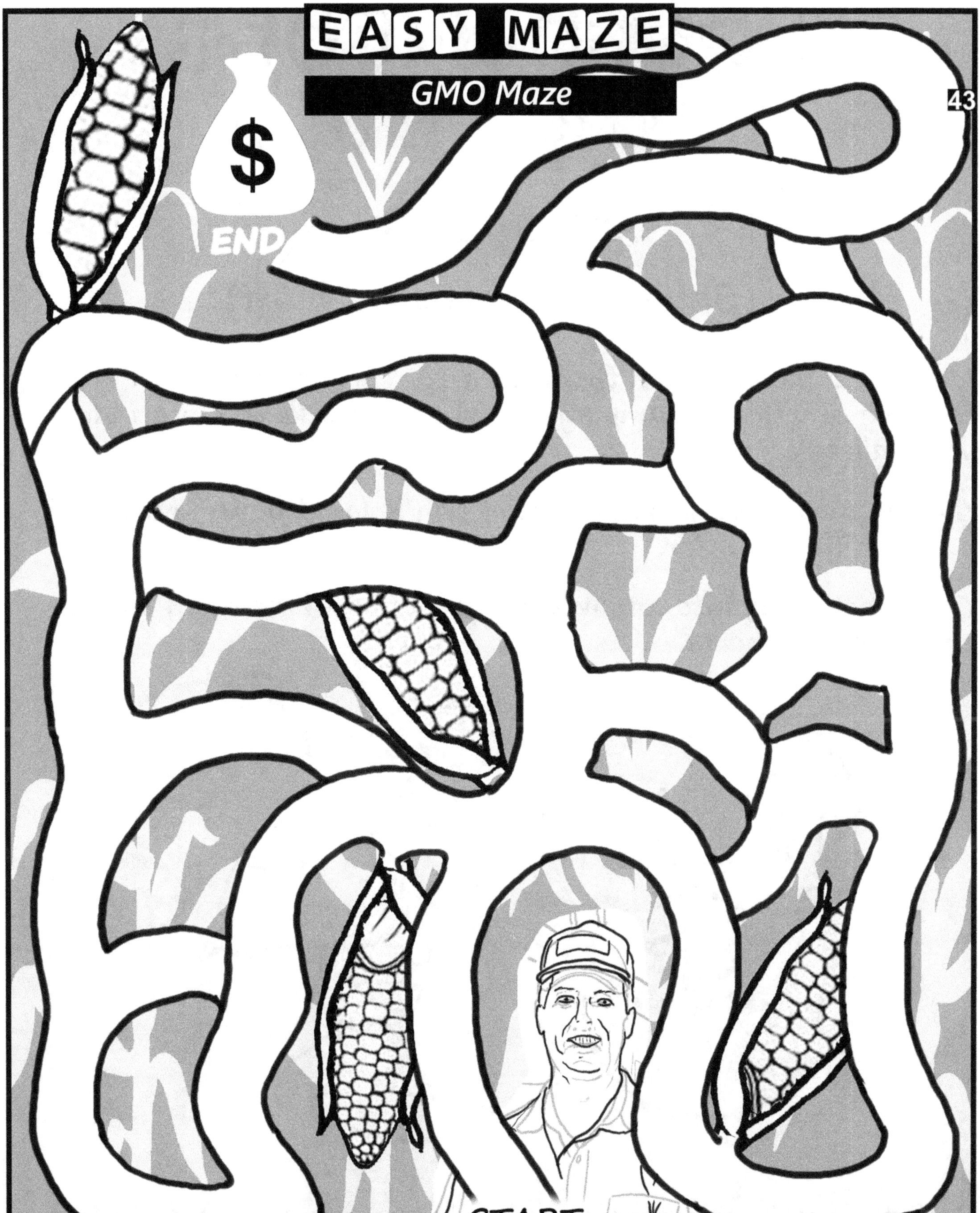

$
END
START

Climb to the heights with a GMO crop,
With this boost to your profits, you won't ever stop.

Chick-ory

(Baby chick treasure hunt)

Ideal for a factory farm themed party. This game is based on the farming practice of dumping all the male chicks into a grinder while alive.

Cut large baby chick shapes out of stiff colored paper, making each a different color.

Chop each chick into a number of pieces and hide them all over the garden or house.

Split the children into pairs, and give them one colored chick to find, showing them a complete example so they know what they are looking for. The first pair to put their chick back together is the winner.

Hatch and Release

With the public outcry over the industrial farming practice of dumping live male chicks into a grinder, the Poultry Public Perception Panel is promoting the "Hatch and Release" method as a more humane solution.

A cushion of air wafts the chicks up the pipe to the release point, where they are free to go wherever they wish.

To avoid waste, those that fall to the receiving area below are respectfully collected and put into a grinder.

Factory Farmer Aptitude Test

Procedure: The test consists of 48 tasks that you will have to rate by how much you would enjoy performing each on a scale of (1) dislike, (2) slightly dislike, (3) neither like or dislike, (4) slightly enjoy, (5) enjoy. The test will take five to ten minutes to complete.

Participation: Your use of this aptitude test should be for educational purposes only. None of the government departments involved in the distribution or grading of this test will respond to public inquiries.

	Dislike		Neutral		Enjoy
Test the quality of parts before shipment	○	○	○	○	○
Perform vivisections	○	○	○	○	○
Operate a beauty parlor or barber shop	○	○	○	○	○
Develop a new medical treatment or medicine	○	○	○	○	○
Foreclose on a property	○	○	○	○	○
Study the structure of the human body	○	○	○	○	○
Teach children to read	○	○	○	○	○
Help elderly people in their daily activities	○	○	○	○	○
Study animal behavior	○	○	○	○	○
Become a prison guard	○	○	○	○	○
Run a toy store	○	○	○	○	○
Run a zoo	○	○	○	○	○
Help people who have problems with animals	○	○	○	○	○
Execute people for the State	○	○	○	○	○
Manage the operations of a hotel	○	○	○	○	○
Operate a grinding machine in a factory	○	○	○	○	○
Direct a play	○	○	○	○	○
Conduct a fugitive search	○	○	○	○	○
Conduct biological research	○	○	○	○	○
Squash insects	○	○	○	○	○
Write a speeding ticket	○	○	○	○	○

	Dislike		Neutral		Enjoy
Conduct a musical choir	○	○	○	○	○
Maintain employee records	○	○	○	○	○
Do volunteer work for a homeless shelter	○	○	○	○	○
Run a sweatshop	○	○	○	○	○
Perform surgical procedures	○	○	○	○	○
Euthanize pets	○	○	○	○	○
Write a song	○	○	○	○	○
Install flooring in houses	○	○	○	○	○
Run over animals	○	○	○	○	○
Supervise the activities of children at a camp	○	○	○	○	○
Perform stunts for a movie or television show	○	○	○	○	○
Work in a biology lab	○	○	○	○	○
Sell chainsaws	○	○	○	○	○
Manage a department within a large company	○	○	○	○	○
Start a fire in a department store	○	○	○	○	○
Investigate fires	○	○	○	○	○
Be an environmental whistleblower	○	○	○	○	○
Play a musical instrument	○	○	○	○	○
Operate a meat grinder	○	○	○	○	○
Handle customers bank transactions	○	○	○	○	○
Design sets for plays	○	○	○	○	○
Demolish houses	○	○	○	○	○
Work on an offshore drilling rig	○	○	○	○	○
Assemble products in a factory	○	○	○	○	○
Study aberrant behavior	○	○	○	○	○
Inspect meat	○	○	○	○	○
Mutilate clothing	○	○	○	○	○

Factory Farm View

"It's all about the lens"

Mark a winding path with string — along the path lay down pieces of corrugated cardboard with factory farm risks written on them, such as:

- Toxic Algal Blooms
- Manure Spills and Fish Kills
- Herbicide Drifts
- Hog Manure Lagoons Overflow in Hurricane
- Antibiotic-resistant Pathogen Pandemic
- Gulf of Mexico Dead Zone
- Toxic Gas Pluming
- Depletion of Aquifers

Then have each child, one at a time, walk along the string, looking through the wrong end of a pair of binoculars.

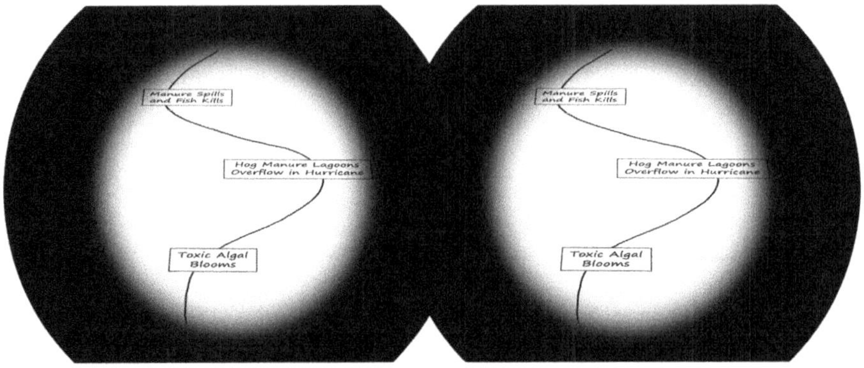
Manure Spills and Fish Kills
Hog Manure Lagoons Overflow in Hurricane
Toxic Algal Blooms
Manure Spills and Fish Kills
Hog Manure Lagoons Overflow in Hurricane
Toxic Algal Blooms

The object is to follow the path of greatest profits, while minimizing the perception of risk. The child who can stay on the path the longest wins.

FARM HARM
COLORING PAGE

COLOR THEM "UNHEALTHY"

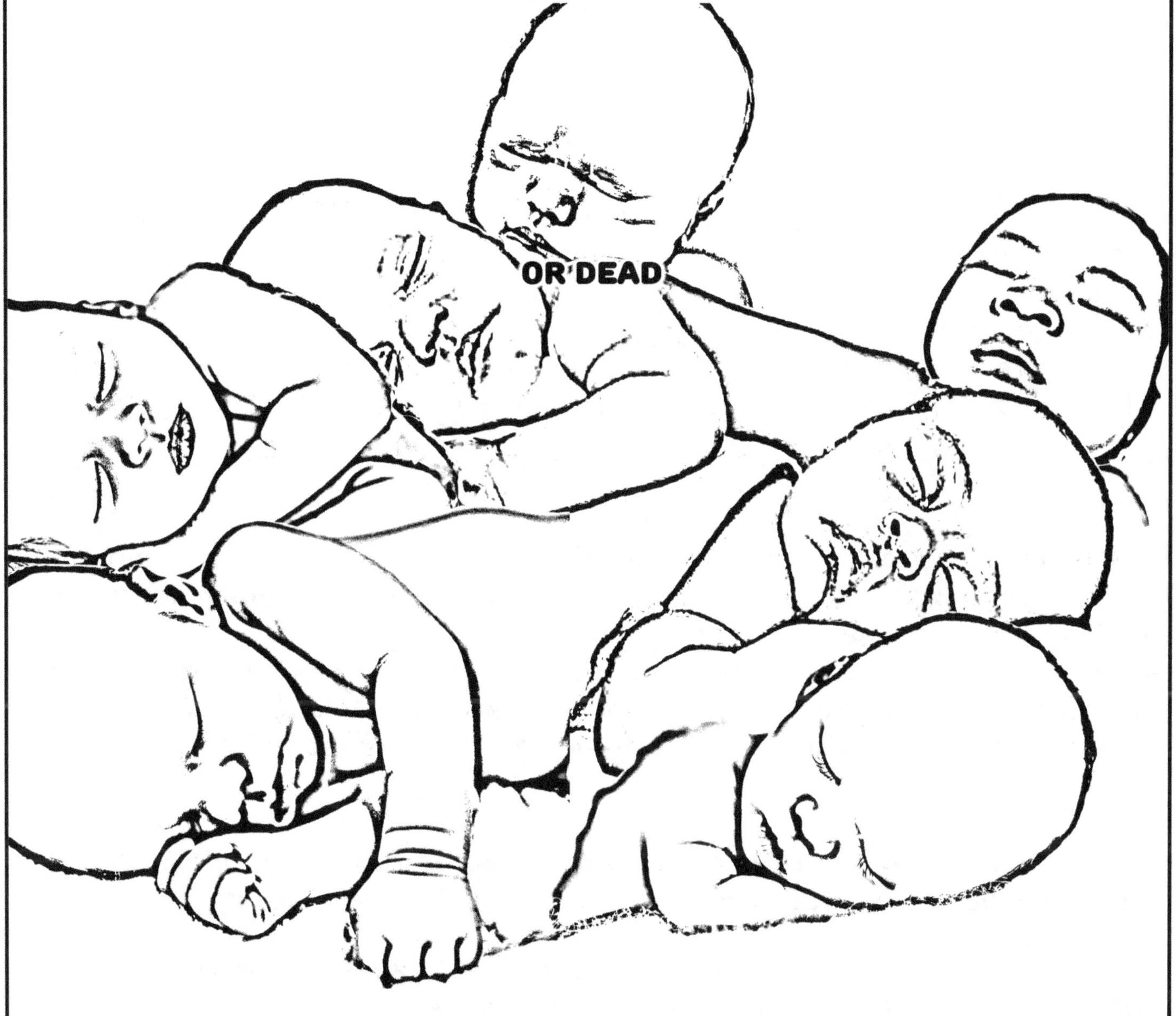
OR DEAD

HOME OWNERS WITH THEIR OWN WELLS ARE SOLELY RESPONSIBLE FOR THE QUALITY AND SAFETY OF THEIR WATER.

Cash Cow Piñata

The game that made  *factory farms.*

You need:

- *2 Paper bags - a large one for the body and a smaller one for the head*
- *4 Toilet paper tubes for legs*
- *Markers and tape, newspaper, and an empty milk carton for the cow's stomach*
- *Play money*
- *Blindfold*
- *Stick or baseball bat*
- *Somewhere to suspend the Cash Cow*
- *Plenty of space [sounds like a factory farm already]*

Making the Cash Cow:

- *Cut the top inch off of the large bag to use for a tail.*
- *Decorate the bags and toilet paper tubes with markers to portray a cute dairy cow.*
- *Stuff the head with crumpled newspaper [to represent permit violation notices], and tape the bag closed.*
- *Fill the body with the play money, more crumpled newspaper violation notices, and a sealed milk container filled with dirt or floor sweepings, and tape that bag closed.*
- *Tape the head, tail, and legs to the body, and get ready for the windfall.*

Play the Cash Cow game:

- *Attach the Cash Cow securely to a tree branch, beam, or any other suitable hanging place, at a height that will allow it to swing freely and where the hitters can reach it with a stick.*
- *Line the hitters [factory farmers] up at least 10 feet or so from the Cash Cow.*
- *Blindfold the first "farmer" and give him or her the stick.*
- *Bring each hitter to within about 3 feet of the Cow and ask them to try to hit it with the aim of breaking it. The hitters could also be turned round a couple of times on the spot so that they are a bit disorientated.*

Important: Don't let the farmers that are waiting in line get too near the hitter, as it can be very dangerous.

- *As soon as the Cash Cow breaks open and the contents shower down, all the farmers can scramble to pick up the goodies.*

The play money can be redeemed for candy and treats [the permit violation notices have no practical value.]

How Politicians Do Math

Political Benefits

The corporate and industrial agribusiness lobby is very important and influential

Selected agricultural figures look good in economic reports

There is a strong and well-funded agricultural media campaign

The public has no knowledge or meaningful participation

Rural people are unimportant

+

$WIN-VOTES!

Externalized Costs

Taxes to subsidize Farmer's School Tax Credit

Taxes to subsidize Farmer's Investment Credit

Taxes to subsidize Agricultural Assessments

Taxes to subsidize Minimum Wages for foreign workers

Taxes to pay for Welfare for unemployed American farm workers

Taxes to subsidize Day Care for foreign farm worker's families

Taxes to pay farmers to encourage pollution reduction practices

Taxes to pay for industrial farming clean-up programs

Taxes to subsidize Crop Insurance

The costs of agricultural activities to outdoor recreation and Quality of Life

The costs of agricultural activities to the environment and Fishing Industry

The costs of agricultural activities to the health of the Rural Community

The costs of agricultural activities to the entire world for methane-fueled Global Warming

+

N/A

Stick it to them!

Pin the Tail on the LEGISLATOR

They're tough to pin down!

Trace or copy the tails and legislators. Cut the them apart, and stick a push pin or thumbtack into the small end of each tail. Fasten either legislator to the wall or a sheet hanging in a doorway and you're ready to play.

Give one tail to every player. Blindfold each player in turn and whirl them around once or twice. Using one hand, they must step forward and pin the tail on the first place they touch.

The player fastening the tail nearest to the proper place is the winner!

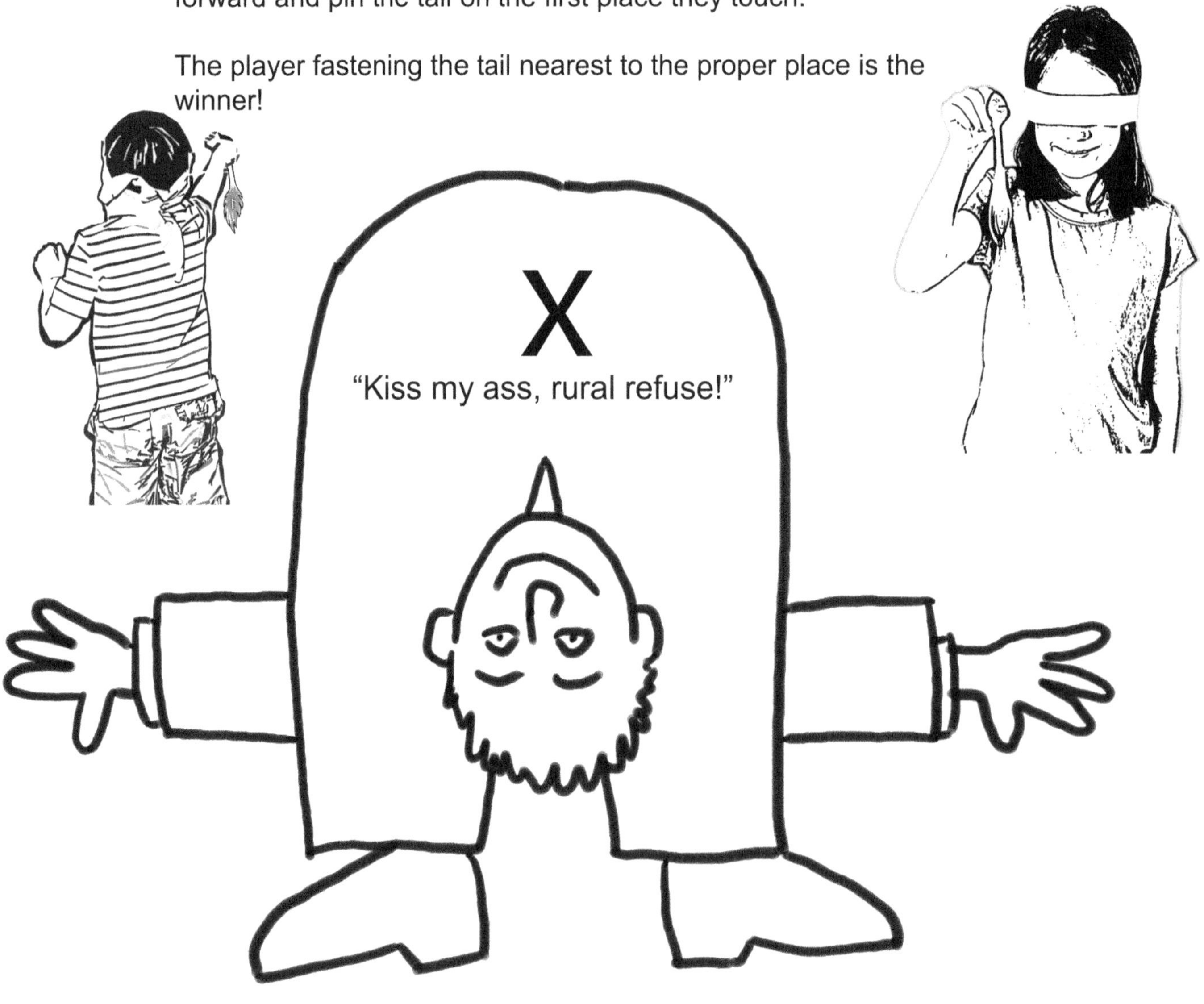
X
"Kiss my ass, rural refuse!"

Q; What do you get when you cross a chicken with a factory farm?
A: Do you want me to tell you now, or after you eat it?

X
“In your dreams, trailer trash!”

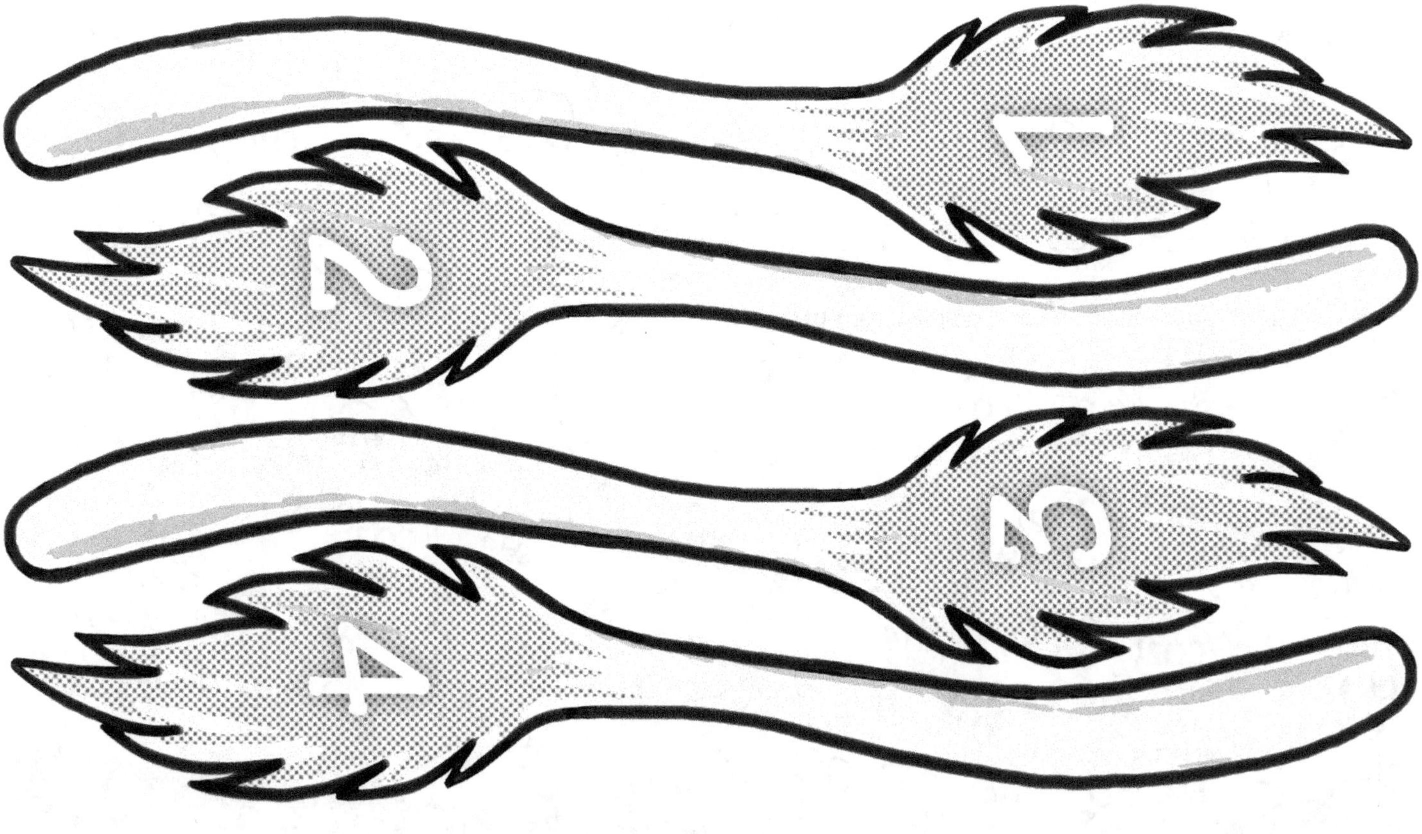
1
2
3
4

Q; What did the herbicide drift say to the neighbor? A: I mist you.

How Many Toxic Gas Plumes

are escaping from this manure pit?

To find out if you counted correctly, mark all the H2S [Hydrogen Sulfide] plumes — The correct number will be revealed.

CH4 [Methane]

NH3 [Ammonia]

CO2 [Carbon Dioxide]

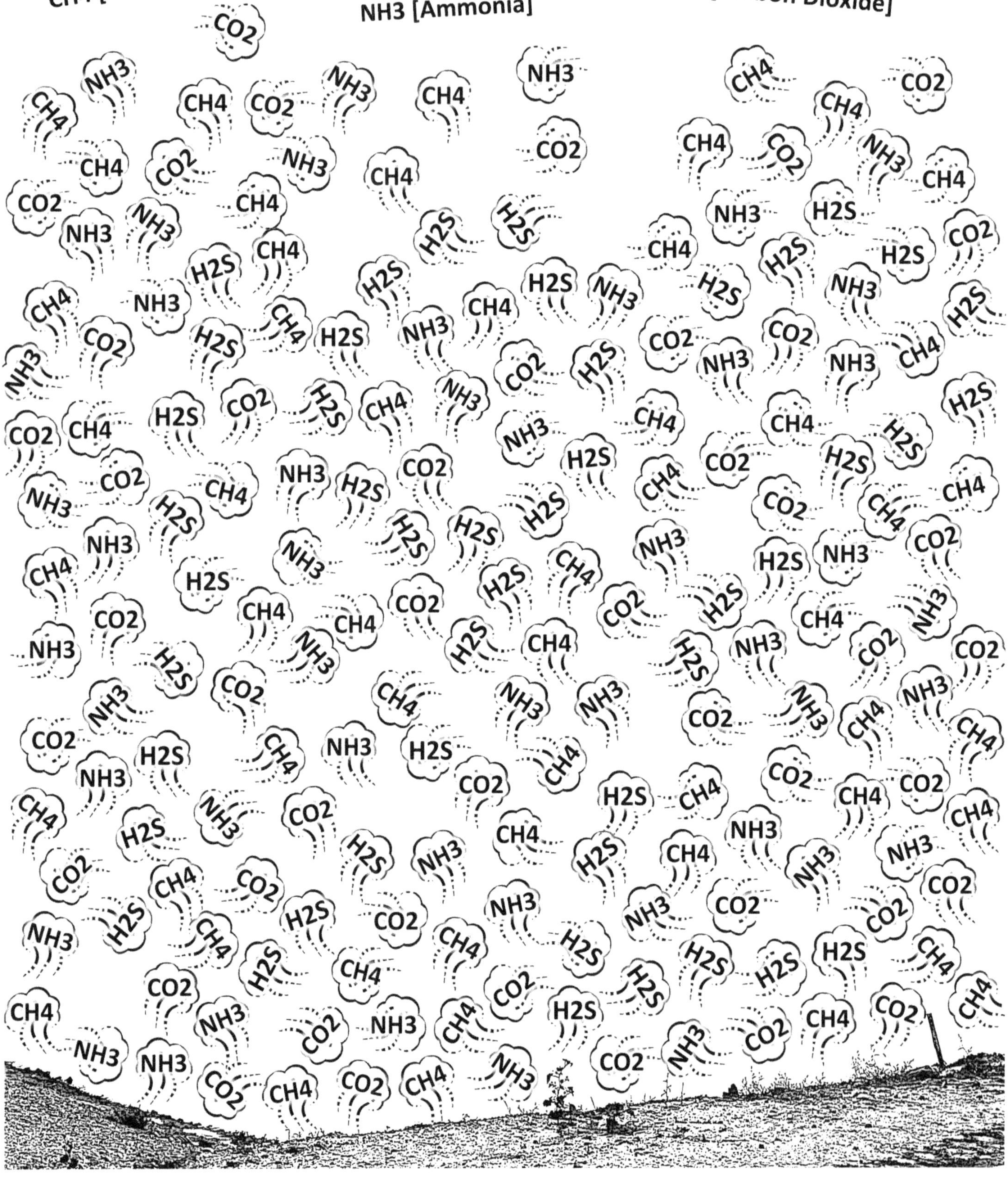

Corporate Pushback Balloon Relay

A simple but instructive party game. The children take the part of Big Ag corporations, their politicians, and funded research. The balloons represent ethical researchers and activists in opposition.

Divide the children into two corporate teams. At the starting signal, the first two contestants in each team run to their basket and grab a balloon. They then sit on it until it pops. As soon as it's popped, they run back to their team and tag the next player.

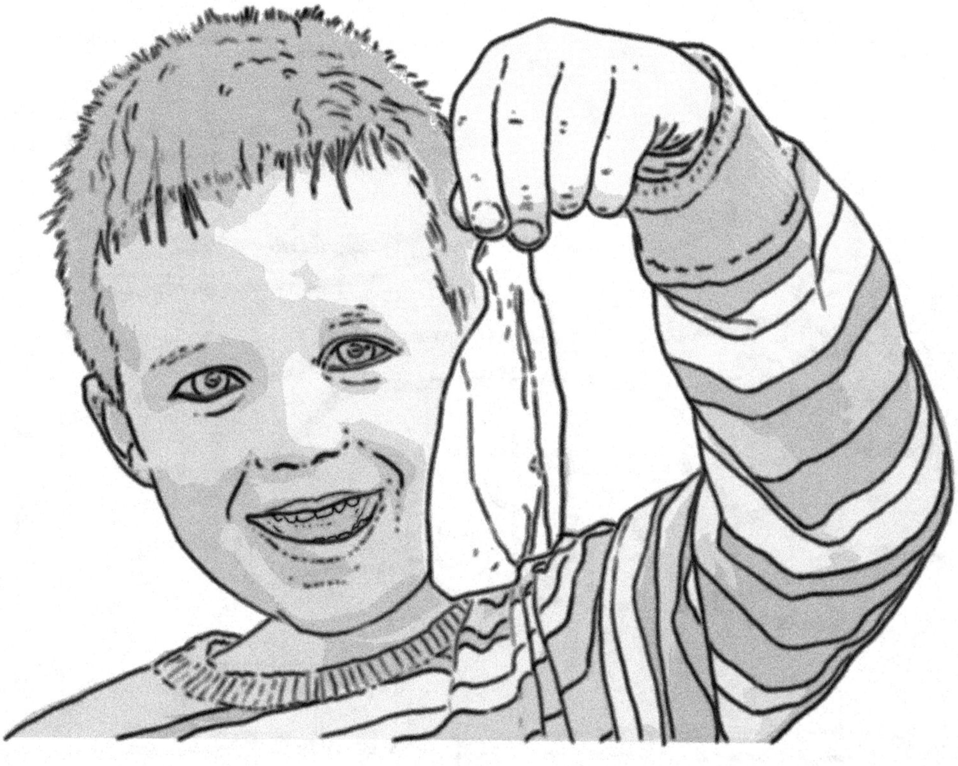

The first corporation to flatten all their opposition wins.

EASY MAZE

Manure Spill Maze

START
Lake Closed
END

Follow the path the manure spill takes,
As it kills off our fish and pollutes our clean lakes.

CON○NEC○T THE DOT○S

Connect numbered the dots for each letter and then hold this page up to a window.

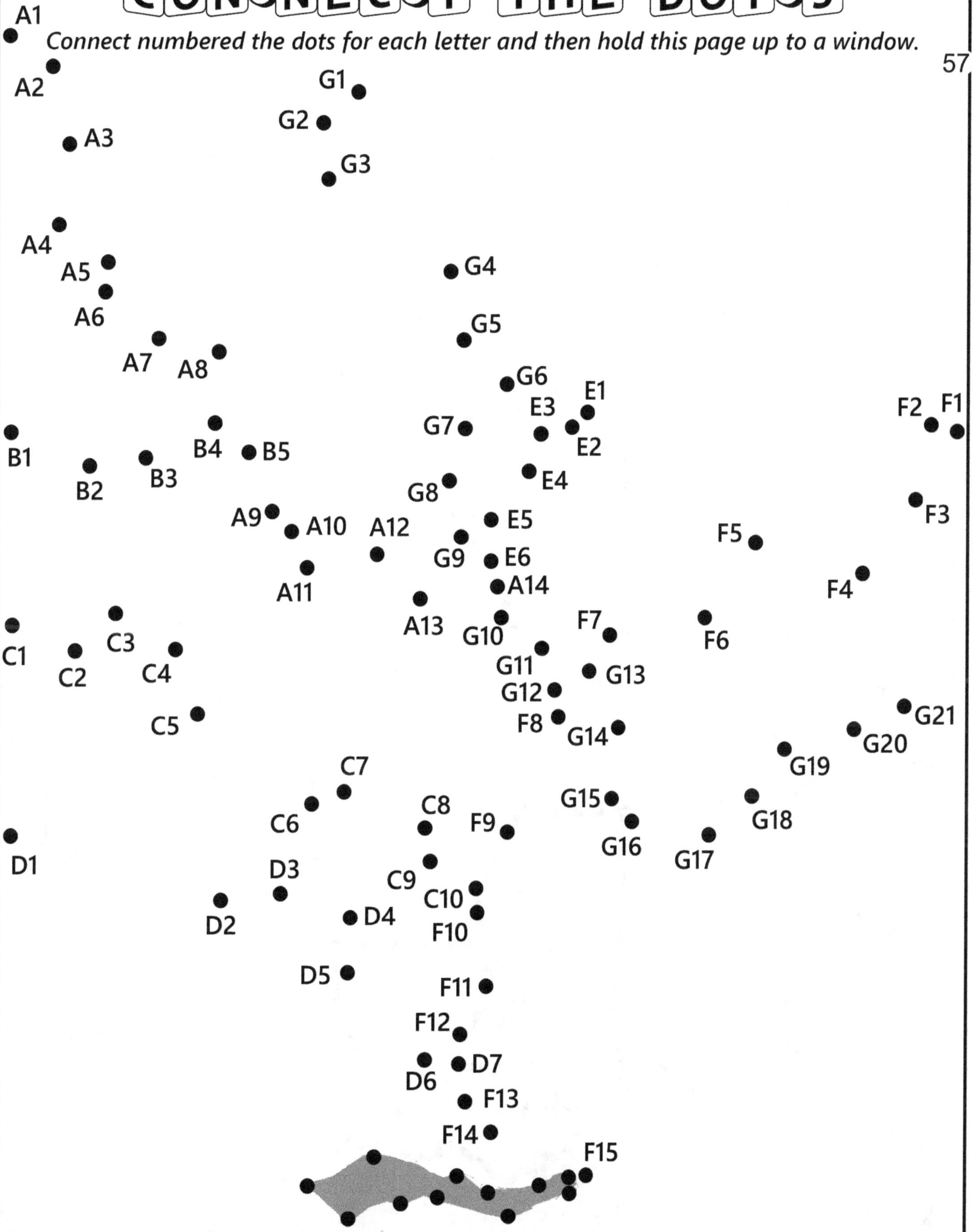
A1
A2
A3
A4
A5
A6
A7
A8
A9
A10
A11
A12
A13
A14
B1
B2
B3
B4
B5
C1
C2
C3
C4
C5
C6
C7
C8
C9
C10
D1
D2
D3
D4
D5
D6
D7
E1
E2
E3
E4
E5
E6
F1
F2
F3
F4
F5
F6
F7
F8
F9
F10
F11
F12
F13
F14
F15
G1
G2
G3
G4
G5
G6
G7
G8
G9
G10
G11
G12
G13
G14
G15
G16
G17
G18
G19
G20
G21

The Dead Zone

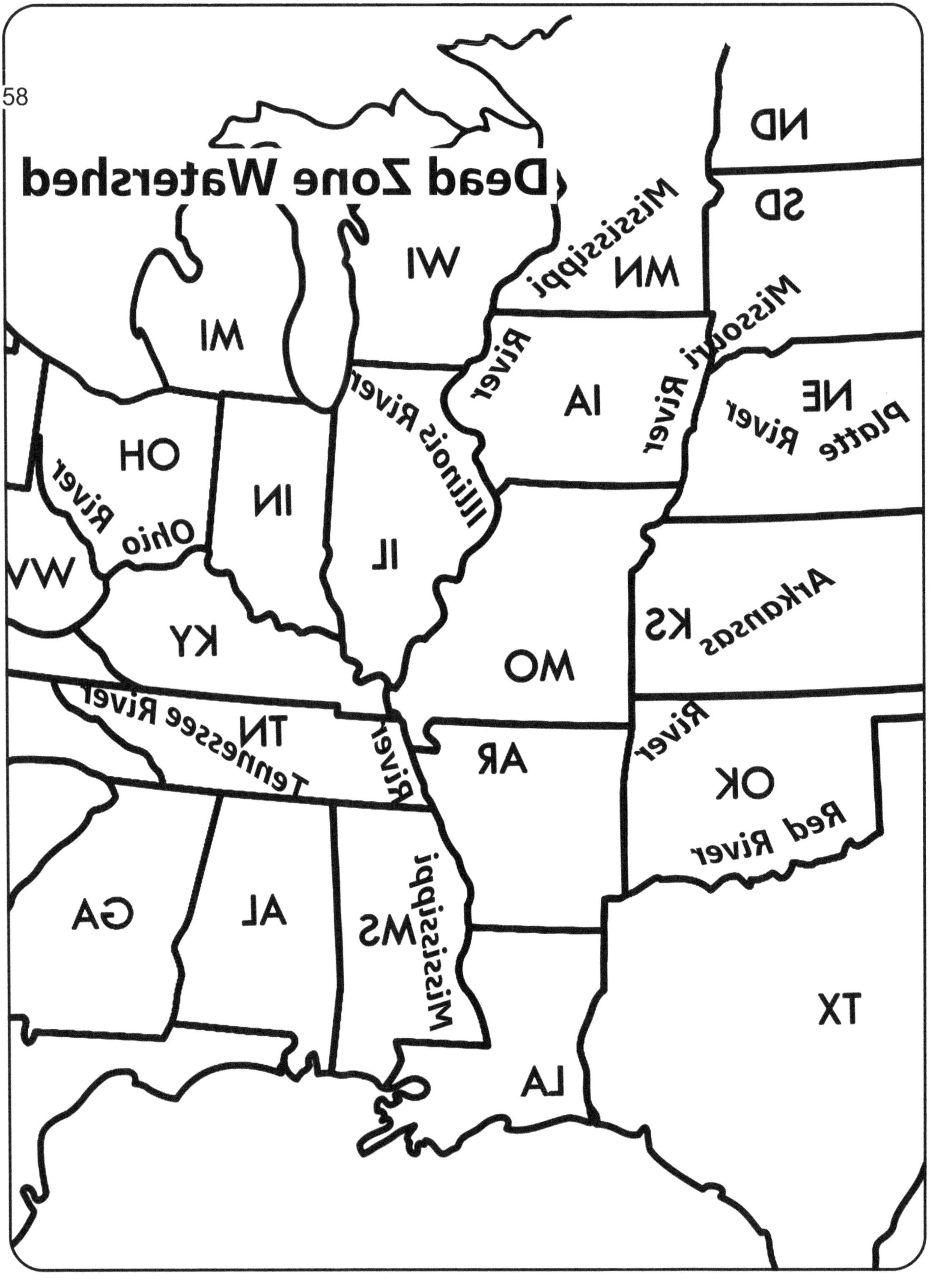
Dead Zone Watershed
ND
SD
Mississippi
MN
WI
MI
Missouri
River
IA
River
NE
Platte
River
Illinois River
OH
River
Ohio
IN
IL
WV
KS
Arkansas
KY
MO
Tennessee River
TN
River
AR
River
OK
Red River
GA
AL
MS
Mississippi
TX
LA

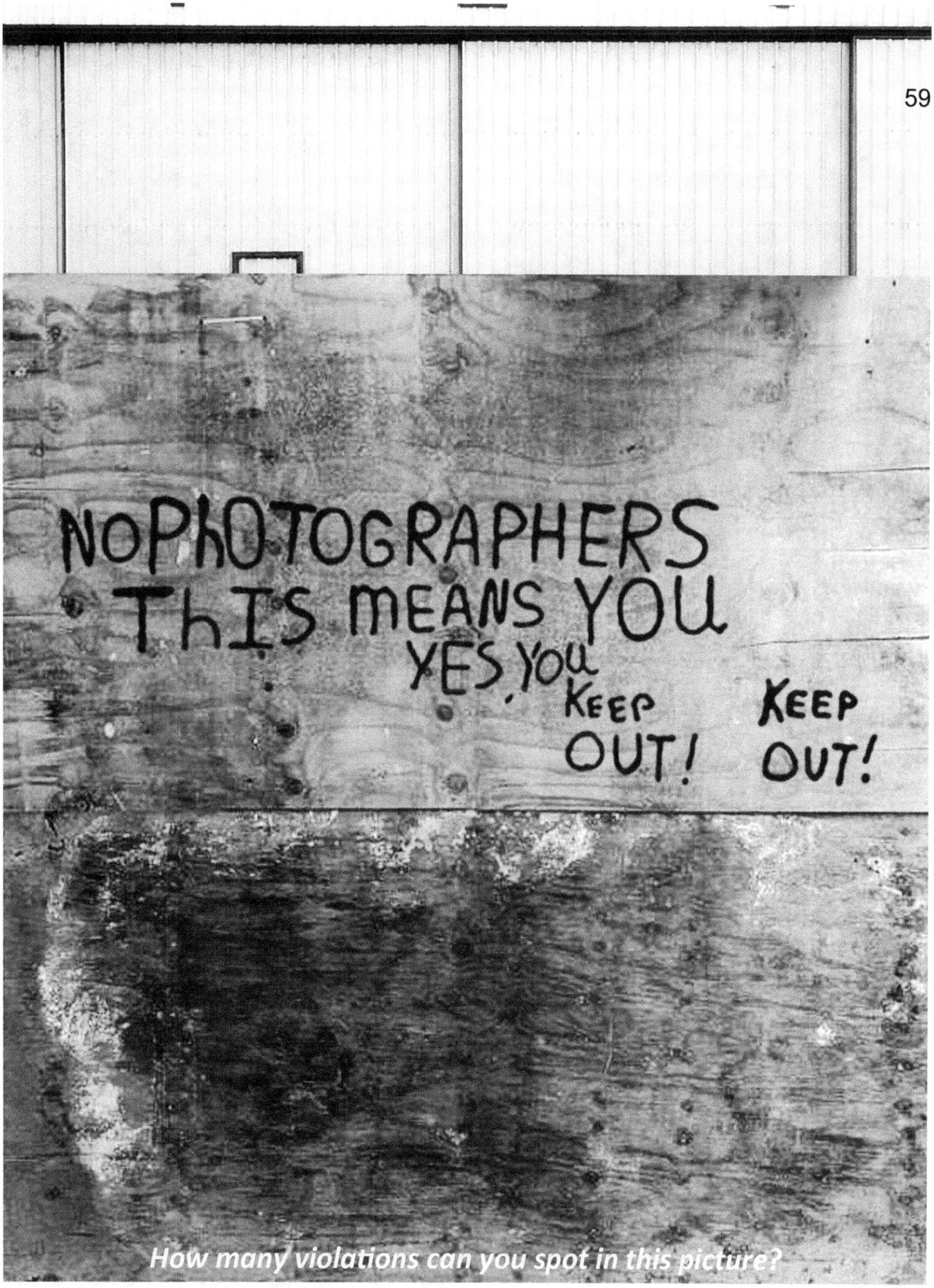
NOPhOTOGRAPHERS
ThIS MEANS YOU
YES, YOU
KEEP OUT!
KEEP OUT!

How many violations can you spot in this picture?

THE
DEAD
ZONE CARD
GAME CRAZE
HUIT MILLE
MILES CARRÉS
8000 SQUARE MILES

Huit Mille Miles Carrés is French for eight-thousand square miles, referring to the size of the "Dead Zone" of hypoxic waters in the Gulf of Mexico that is created by nutrient pollution.

Game Overview:

You are one of up to 3 players attempting to pollute a specific square mileage while dealing with regulatory hazards you encounter along the way. There are 106 cards, including hazards, Dead Zone areas, remedies, and four immunity cards.

Setup:

Each player will need a play area for their area, battle, regulations, and immunity piles. You'll want the Dead Zone area pile to be ordered numerically to assist in tracking area polluted.

Each player is dealt a single card, beginning with the left of the dealer, and dealing continues to six cards. Play begins to the dealer's left. Each player must draw one card at the beginning of their turn and play or discard one card to end their turn. Play continues until one player reaches 8,000 square miles exactly (you can't go over), or the final card is played after the draw pile has expired.

Playing the game:

After drawing one card during your turn, you may play one of the following:

Area Cards: These represent the total Dead Zone area you have amassed during the game, with the overall goal being to reach 8,000 square miles first. You may play any combination of area cards, but you may play only two 1600 square miles cards and you must have a *Permit Approval* card on your regulations pile or *Court Reverses Decision* card on your immunity pile to place any area cards.

200
400
600
800
1600

Hazard Cards: These allow you to disrupt your opponent's movement by playing a hazard onto your opponent's battle pile. The hazards are *TDML restrictions, EPA oversight, CAFO opposition,* and *Conservation Committee*. Each hazard has its corresponding remedy card that a player may play over the hazard on their own turn. *Permit Approval*, or *Court Reverses Decision* cards must be placed in order to continue placing Zone area cards, even if a hazard has been remedied.

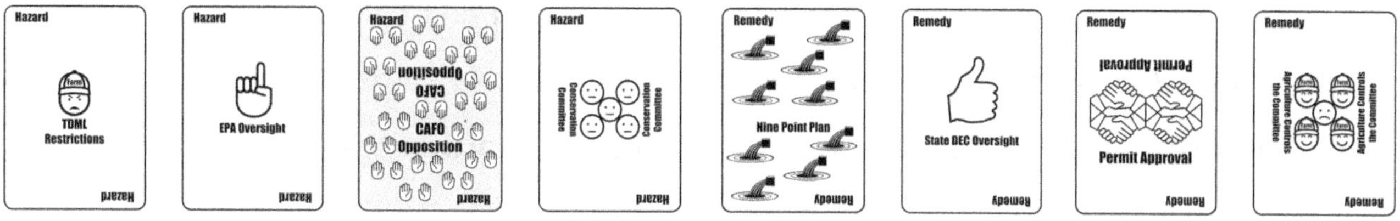
Hazard
TDML Restrictions
Hazard
EPA Oversight
Hazard
CAFO Opposition
Hazard
Remedy
Nine Point Plan
Remedy
State DEC Oversight
Remedy
Permit Approval
Remedy

Regulations Cards: These may be placed onto your opponents's regulations pile to restrict that player from using any area cards. There is a corresponding *Voluntary Guidelines* card that a player may use to remove this restriction during their own turn.

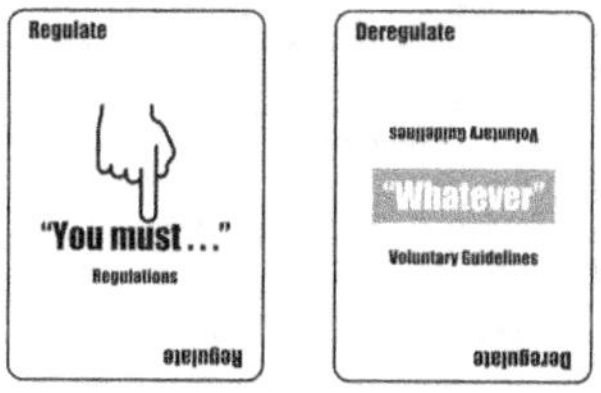
Regulate
"You must..."
Regulations
Deregulate
"Whatever"
Voluntary Guidelines

Immunity Cards: There are four cards that grant immunity to corresponding hazards. They may all be played during your turn or used as a Coup Fourré. If it is played on your turn you may remove the corresponding hazard it protects against (if one is present) and take another turn. If you decide to Coup Fourré, your opponent must play a hazard that your immunity card (still in your hand) protects against. Before the next card is drawn, you must call Coup Fourré and place your immunity card in the immunity pile on its side. This earns you bonus points at the end of the round. You also take your turn immediately, ending the opponent's current turn and skipping other players if necessary.

Immunity
Governor Steps In
Immunity
Industry Funded Research
Conflicting Research
Immunity
Deadline Pushed
Back
Immunity
Courts Overturn Decisions

Whenever a hazard is played, any player holding the corresponding immunity card may immediately play it and declare a Coupe Fourré.

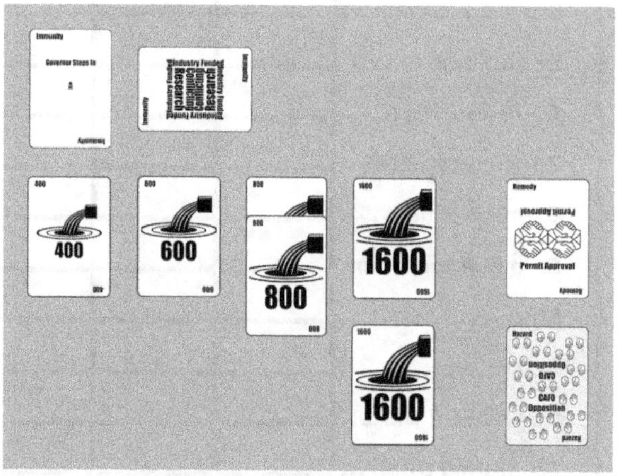
400
600
800
1600
1600
Example of a play area

Scoring and Winning the Game:

At the end of the round each player tallies their score, with the goal being to reach 40,000 points first. Dead Zone area cards translate 1:1 to your score. If you pollute 800 square miles, you get 800 points.

Bonus points are awarded additively under the following conditions:

Complete 8000 square mile Dead Zone = 3200

Each Immunity card in your safety pile = 800

All four Immunity cards in your safety pile = 2400

Immunity card as Coup Fourré = 2400

Delayed Action (complete 8000 square mile Dead Zone after last card is drawn) = 2400

Safe Nutrient Plan (you did not play any 1600 square mile area cards) = 2400

Shut out (your opponent did not play any Dead Zone area cards) = 4000

Make copies and cut out score cards as needed.

HUIT MILLE MILES CARRÉS Score Card

Player or team			
Brought forward			
Dead Zone area			
Immunities			
Coup-fourrés			
Zone completed			
Delayed action			
Safe Nutrient Plan			
Shut-out			
Extension			
TOTAL FOR DEAL			
Dead Zone area			
Immunities			
Coup-fourrés			
Zone completed			
Delayed action			
Safe Nutrient Plan			
Shut-out			
Extension			
TOTAL FOR DEAL			
Combined total			

HUIT MILLE MILES CARRÉS Score Card

Player or team			
Brought forward			
Dead Zone area			
Immunities			
Coup-fourrés			
Zone completed			
Delayed action			
Safe Nutrient Plan			
Shut-out			
Extension			
TOTAL FOR DEAL			
Dead Zone area			
Immunities			
Coup-fourrés			
Zone completed			
Delayed action			
Safe Nutrient Plan			
Shut-out			
Extension			
TOTAL FOR DEAL			
Combined total			

800
800
800

Remedy
State DEC Oversight
Remedy

Remedy
Agriculture Controls the Committee
Farm
Agriculture Controls the Committee
Remedy

Guide for **Huit Mille Miles Carrés**

The deck should consist of 106 cards in these combinations.

Attacks (Cartes d'attaque)	**Parades** (Cartes de défense)	**Immunity** (Cartes de securité et de coup-fourré)
3 TDML restrictions	6 Nine Point Plan	1 Conflicting research
3 EPA oversight	6 State DEC oversight	1 Deadlines pushed back
3 Conservation Committee	6 Agriculture Controls Committee	1 Governor Steps In
4 Regulations	6 Voluntary Guidelines	1 Courts overturn decisions
5 CAFO opposition	14 Permit Approval	

STEPS (cartes de la région) in square miles

Four 1600—Twelve 800—Ten 600—Ten 400—Ten 200

800
800
800

Deregulate
Voluntary Guidelines
"Whatever"
Voluntary Guidelines
Deregulate

Remedy
Nine Point Plan
Remedy

HUIT MILLE MILES CARRÉS
HUIT MILLE MILES CARRÉS

Make 6 copies of this page and cut out cards on the dotted lines. [Refer to guide card for the number of cards used in playing.]

Make 5 copies of this page and cut out cards on the dotted lines. [Refer to guide card for the number of cards used in playing.]

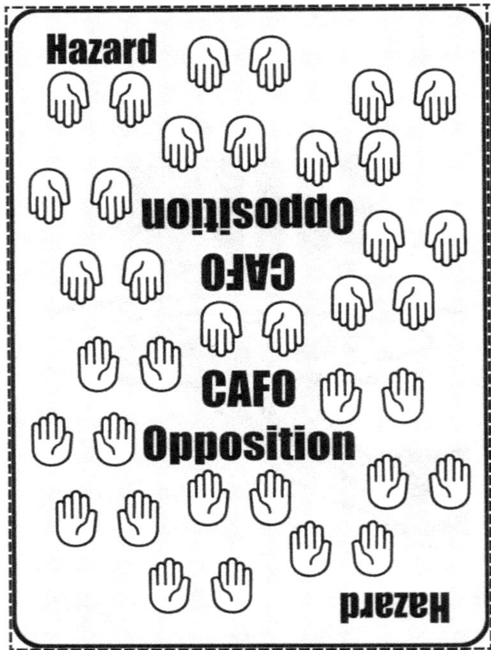
Hazard
Opposition
CAFO
CAFO
Opposition
Hazard

600
600
600

400
400
400

200
200
200

Regulate
"You must . . ."
Regulations
Regulate

600
600
600

400
400
400

200
200
200

1600

1600

1600

1600

1600

1600

Immunity

Deadline
Pushed
Back

Immunity

Immunity

Governor Steps In

Immunity

1600

1600

1600

1600

1600

1600

Immunity

Courts Overturn Decisions

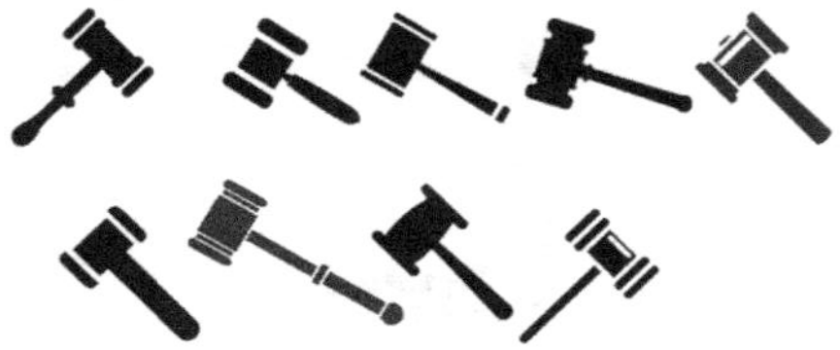

Courts Overturn Decisions

Immunity

Immunity

Industry Funded
Research
Conflicting
Conflicting
Research
Industry Funded
Industry Funded
Industry Funded

Immunity

Make a copy of this page and cut out cards on the dotted lines. [Refer to guide card for the number of cards used in playing.]

Make 3 copies of this page and cut out cards on the dotted lines. [Refer to guide card for the number of cards used in playing.]

Remedy
Permit Approval
Permit Approval
Remedy

Remedy
Permit Approval
Permit Approval
Remedy

Hazard
EPA Oversight
Hazard

Hazard
Conservation
Committee
Conservation
Committee
Hazard

Remedy
Permit Approval
Permit Approval
Remedy

Remedy
Permit Approval
Permit Approval
Remedy

Remedy
Permit Approval
Permit Approval
Remedy

Hazard
Farm
TDML
Restrictions
Hazard

CAFO Sing-a-long

Like "100 Bottles of Beer" except the numbers keep increasing
[There's no music, but rural people are familiar with the tune.]

100 GMO Cows in a shed, 100 GMO cows.
They made more soon and built a lagoon,
200 GMO cows in a shed.

200 GMO cows in a shed, 200 GMO cows.
They made more soon and built a lagoon,
300 GMO cows in a shed.

300 GMO cows in a shed, 300 GMO cows.
They made more soon and built a lagoon,
400 GMO cows in a shed.

400 GMO cows in a shed, 400 GMO cows.
They made more soon and built a lagoon,
500 GMO cows in a shed.

500 GMO cows in a shed, 500 GMO cows.
They made more soon and built a lagoon,
600 GMO cows in a shed.

600 GMO cows in a shed, 600 GMO cows.
They made more soon and built a lagoon,
700 GMO cows in a shed.

700 GMO cows in a shed, 700 GMO cows.
They made more soon and built a lagoon,
800 GMO cows in a shed.

And so on, and so on . . .

Child Safety Tips — *Manure Lagoons*

Your child is always at risk from toxic manure lagoons — even if they can't see or smell them. How can you make sure they are safe? Teach your children:

FARM HARM

1 What do you mean by toxic manure lagoons?

Manure lagoons have been shown to harbor and emit substances which can cause severe health effects to children or even death.

2 Who can they go to when they need help?

Local Authorities **DEC** **USDA**

3 Over 400 VOCs and toxic gases including:

- **Hydrogen Sulfide**
- **Methane**
- **Ammonia**

4 More than 150 dangerous pathogens including:

- **E. coli**
- **Cryptosporidium**
- **Anthrax [Bacillus anthracis]**
- **Leptospira pomona**
- **Listeria monocytogenes**
- **Salmonella**
- **Clostridium tetani**
- **Histoplasma capsulatum**
- **Microsporum and Trichophyton Ringworm**
- **Giardia lamblia**
- **Cryptosporidium**
- **Pfiesteria piscicida**

5 Growth hormones, heavy metals, antimicrobials and antibiotics:

- **rBST, estrogen, testosterone**
- **Arsenic, copper, selenium, zinc, cadmium, molybdenum, nickel, lead, iron,manganese, aluminum, and boron**
- **Antimicrobials and antibiotics administered at sub-therapeutic levels produce antibiotic resistant bacteria**

6 What to teach your child:

Do not touch or step in water around a farm - it may be bad.

Always wear a particle mask or respirator when playing outside.

Stay away from manure lagoons — or you may die.

NO!

Parents

Agricultural studies are intended to protect the profits of farmers, not the health of your children.

Teach them about Farm Harm.

FARM HARM

COLORING PAGE

Hydrogen Sulfide Methane Ammonia

AND OVER 400 VOCs AND TOXIC GASES

DANGER
LIQUID MANURE STORAGE
PELIGRO
ALMACENAJE DE ESTIÉRCOL LÍQUIDO

Antibiotics Estrogen Testosterone rBST Antimicrobials Arsenic Nickel Anthrax Salmonella Cadmium E.coli Cryposporidium Leptospira pomona Molybdenum Pfiesteria piscicida Microsporum/Trichophyton ringworm Giardua lamblia Selenium Bacteria Listeria monocytogenes Lead Iron Boron Clostridium tetani Copper Histoplasma capsulatum Aluminum and hundreds of dangerous pathogens, gorwth hormones, heavy metals, and antibiotic-resistant bacteria

LIQUID MANURE PIT

TONGUE TWISTERS

How many subsidies should a
selfish sucker seek if a selfish
sucker should seek subsidies?

Phosphorus should so silt shut
salty seashore shelves, sluggish
salmon shall shun sandy slits,
singly seeking seasonal shallows.

Soft snow stagnant spreading
slurry streams sluggish shallow
sheets seeping sliding sweeping
steep slick slopes slips swiftly
streaking the straight still shore.

Ag Uncertainty Principle

Agricultural pollution
and the accountability
of agriculture
cannot be measured exactly,
at the same time,
even in theory.

The very concepts
of agricultural pollution
and accountability
together,
in fact,
have no meaning in politics.

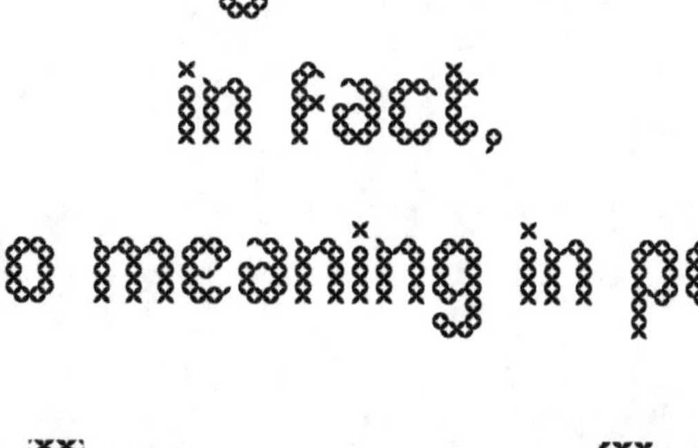

Agri-mandias

I met a traveler in a barren land
Who said: two vast rows of concrete blocks
Stand in the desert. Near them in the sand,
half sunk, a line of shattered silos lie, whose brown
And rusted shapes and stamp of cold command
Tell that its makers well knew that arrogance
Which yet survives, stamped on these lifeless things,
The hands that made them and the hearts that fled:
And on a crumbling wall these words appear
"We feed the World!"
Nothing beside remains. Round the decay
Of that colossal folly, boundless and bare
The waterless sands stretch far away.

(With apologies to Percy Bysshe Shelly, and none to industrial farming.)

Follow the Regulator

Get the children to follow someone playing the part of an Agricultural Regulator, starting with an adult demonstrating to get the ball rolling.

They have to copy whatever the regulator does as they move around, such as:

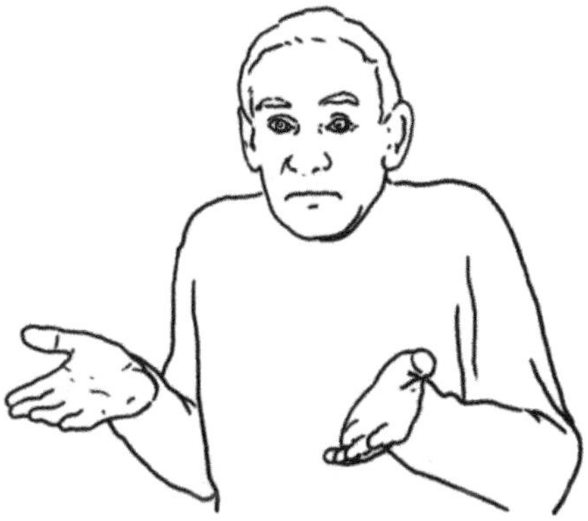

The Shrug
"There's nothing we can do about that"

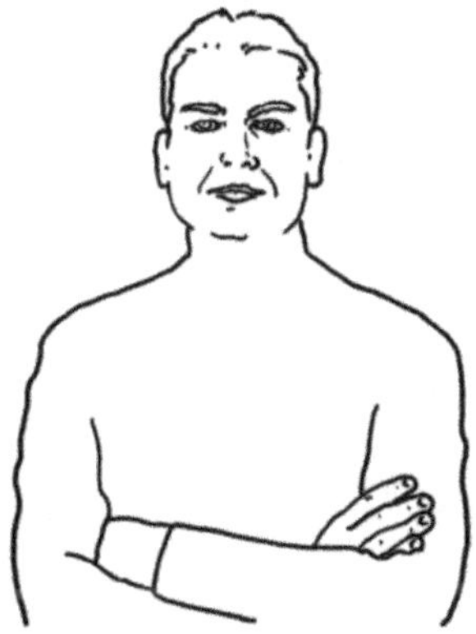

Arms Crossed
"You can't prove that"

Puzzled
"I don't understand the question"

Pushing Away
"You're a troublemaker"

Pointing
"You better stop saying that"

Thoughtful
(They'll think I'm considering it)

All these gestures can also be used in "The Regulator" line dance.

Rural Sorrow
HOPSCOTCH

Lay out the numbers as shown in either of the "courts" pictured by scratching in the dirt or drawing with chalk on pavement.

Toss the marker [a stone, coin or bean bag] into square one. The marker must land completely in the square without touching the line.

9
10
8
7
5
6
4
2
3
1

Skip that square and hop on all the numbers using one foot for single number squares and both feet for doubles, turn and hop back, stop in the square before your marker, bend over and pick it up on the way.

[Neutral squares marked "Safe', "Home", or "Rest" may be hopped through in any manner without penalty.]

If you overthrow the marker, lose your balance, step on a line, or into a square with another player's marker, your turn ends. Upon successfully completing the sequence, the player continues the turn by tossing the marker into square number two, and repeating the pattern.

Players begin their turns where they last left off. The first player to complete one course for every numbered square on the court wins.

The poem on the following page is based on a traditional children's nursery rhyme about magpies. According to an old superstition, the number of magpies one sees determines if one will have bad luck.

Q: According to government studies, what is the effect of factory farms on the health and well-being of their neighbors? A: There are no such studies.

REST
9
10
8
7
5
6
4
2
3
1

One is for sorrow at the passing of your wife

Two is for the joy that has been taken from
your life

Three is for your girl with early puberty
onset

Four is for the teenage boy whose loss
you still regret

Five is for the silver coins corruption counts
as gain

Six is for their love of gold – the source of
all your pain

Seven for the secrets that are never to be told

Eight is for the fish that died, one-hundred
thousand fold

Nine is for the infant child, whose cold
still face you kiss

Ten is for the birds, whose song you're told
you'll never miss

Q: What do you do when a factory farm pollutes your well? A: That's *your* problem.

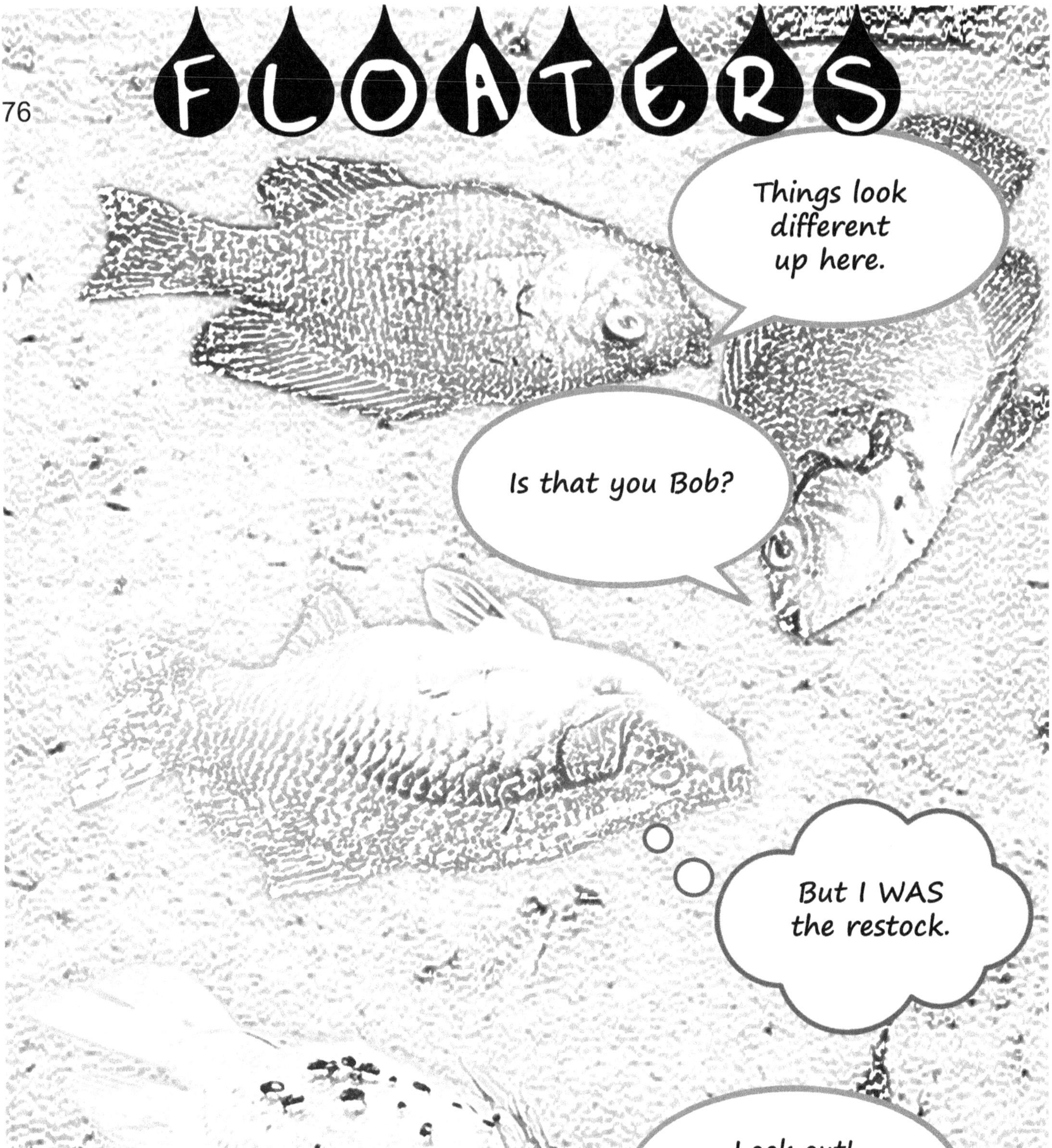
FLOATERS
Things look different up here.
Is that you Bob?
But I WAS the restock.
Look out! Here come the blowflies.

Knock, knock, - Who's there? - It's Bill. - Bill who? - Not Bill, *SPILL* — manure's spilled into the river again!

Anonymous Ag Survey

The reality of living in-and-around a factory farm is very different from the one portrayed in their TV ads and media handouts. Ethical and environmental considerations often take a back seat to profit and convenience. The only way to find out what's really happening in agriculture is with an anonymous survey.

Below are a few questions that can be used in creating your own survey for a class project.

Question	Answer
Do you dump toxic chemicals and animal carcasses in your back lot?	Yes ○ No ○
Does your farm investment tax credit include an in-ground pool or a yacht?	Yes ○ No ○
Have you ever used "Nuisance Law" protection to get back at a neighbor?	Yes ○ No ○
Do you really follow your nutrient management plan?	Yes ○ No ○
How many times has the state DEC covered up for you?	0 ○ 1 ○ 2 ○ 3 ○ 4 ○ 5 ○ 6 ○ 7 ○ 8 ○ 9 ○ 10+ ○
Do you ever wish you had a whip to make those lazy workers move faster?	Yes ○ No ○

Q: If Agriculture is such a great economic asset to our state, why do we have to subsidize their crops and their livestock and their buildings and their machinery and their gas tax and their school tax and their workers pay and their environmental remediation for them to be economically viable?

ORIGAMI HOGS

While there are many better looking origami hogs, they take more time to fold and require many more steps — and with factory farms, it's not about the quality, it's about the quantity. You and a few friends could fill a shed in no time.

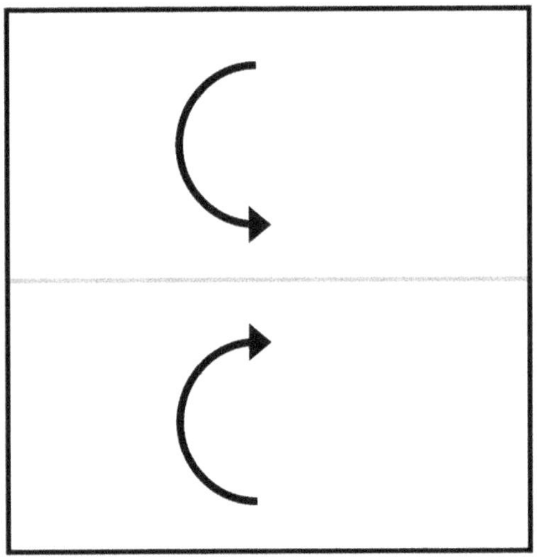

1. Copy and cut out the square design in the lower right hand corner of the facing page. Then turn to the blank side and fold in half and unfold. Fold at the dotted lines so the edges meet at the center.

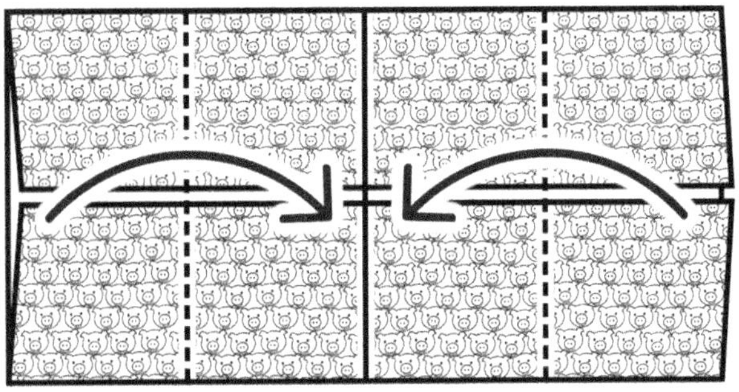

2. Fold at the dotted lines as shown to make creases. Then unfold.

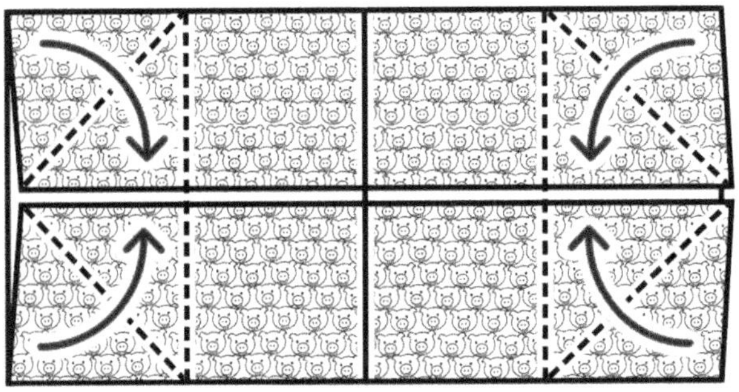

3. Fold corners in at the dotted lines as shown. Then unfold.

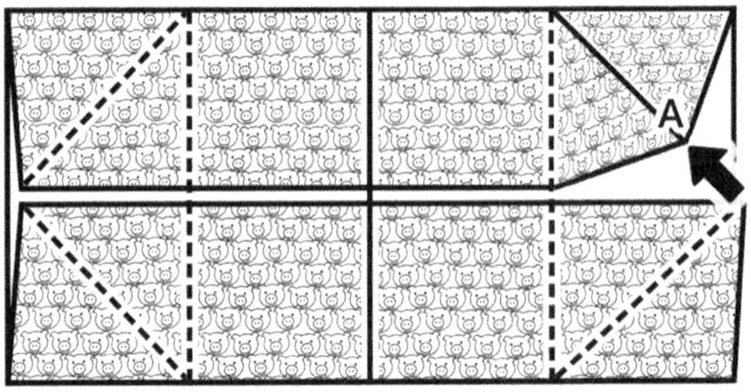
A

4. Lift up corner from underneath at point A [moving the crease upward] and bringing that point to the center as shown below.

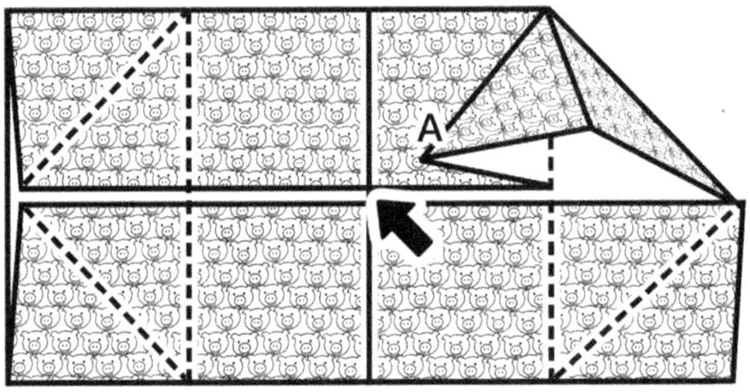
A

5. Flatten. Do the same at the other three corners so it like the diagram below.

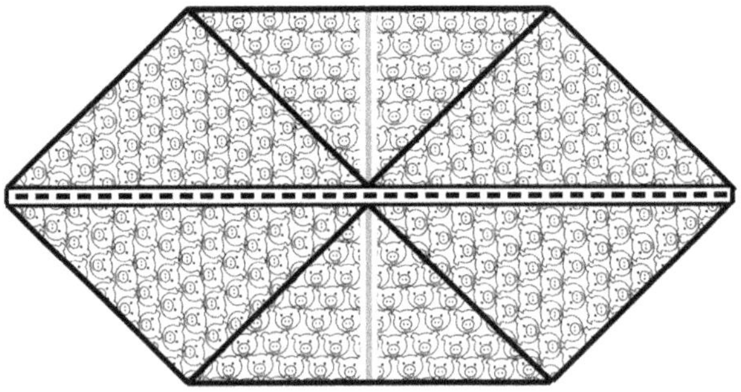

6. Fold backwards at the dotted line.

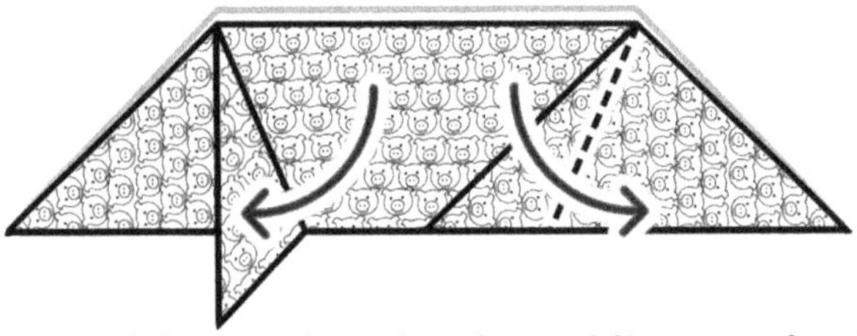

7. Fold forward at the dotted lines on both sides to make the four legs.

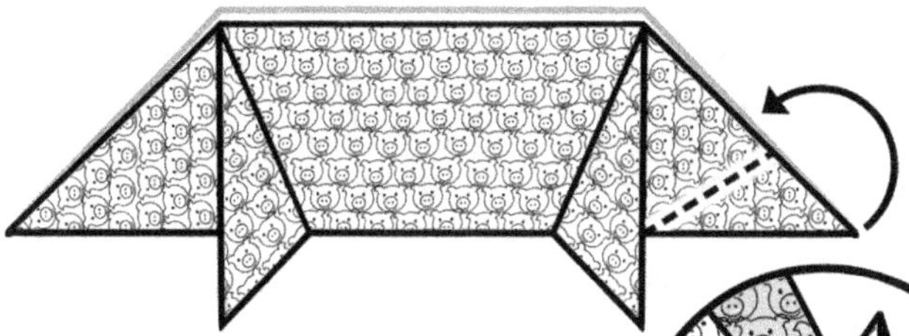

8. Fold inside at the dotted line for the tail.

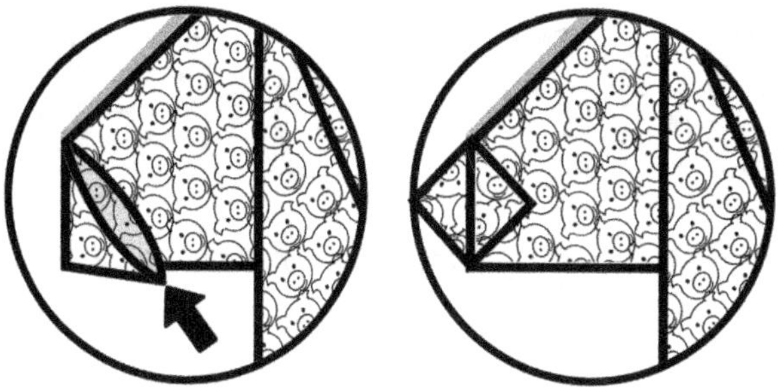

10. Push open the pocket and flatten to make a square for the snout.

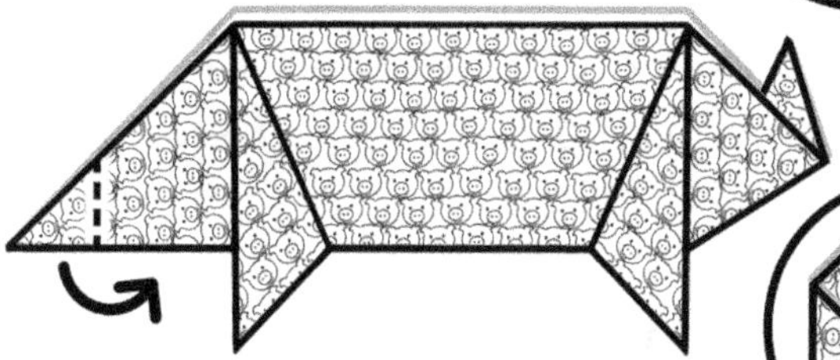

9. Fold forward at the dotted line.

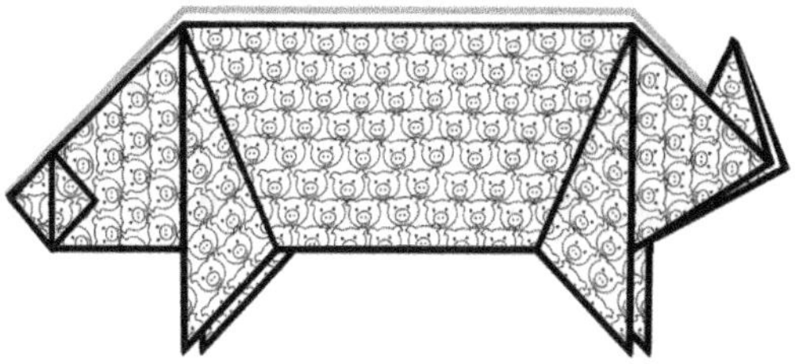

11. Finished origami hog.

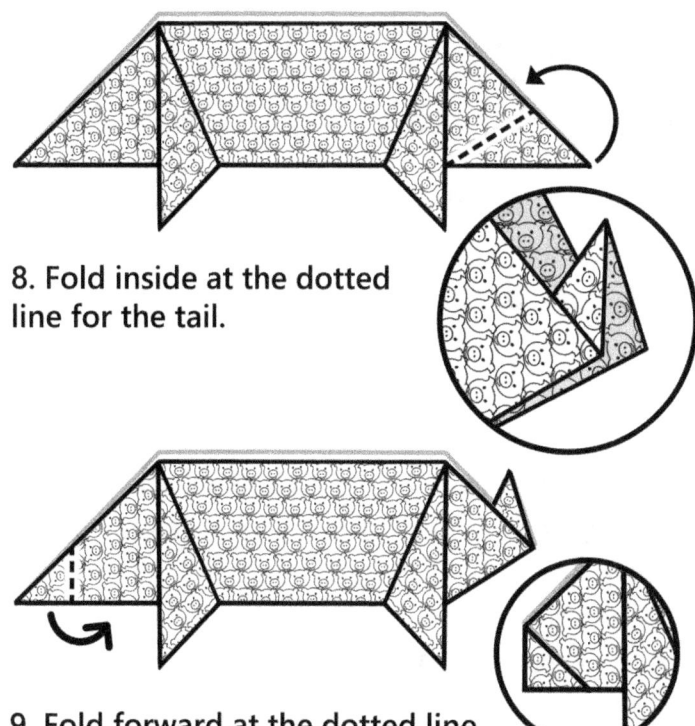

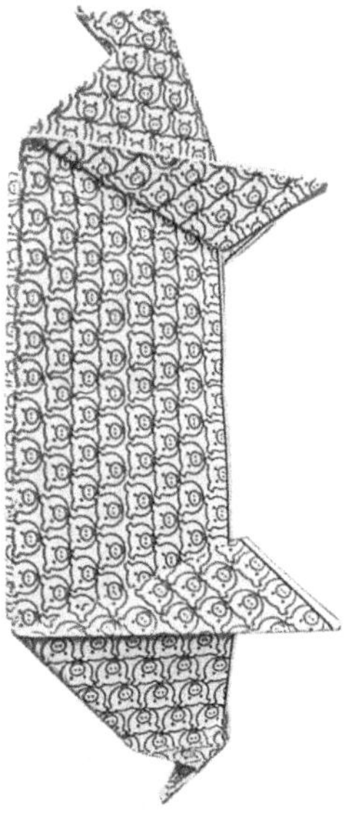

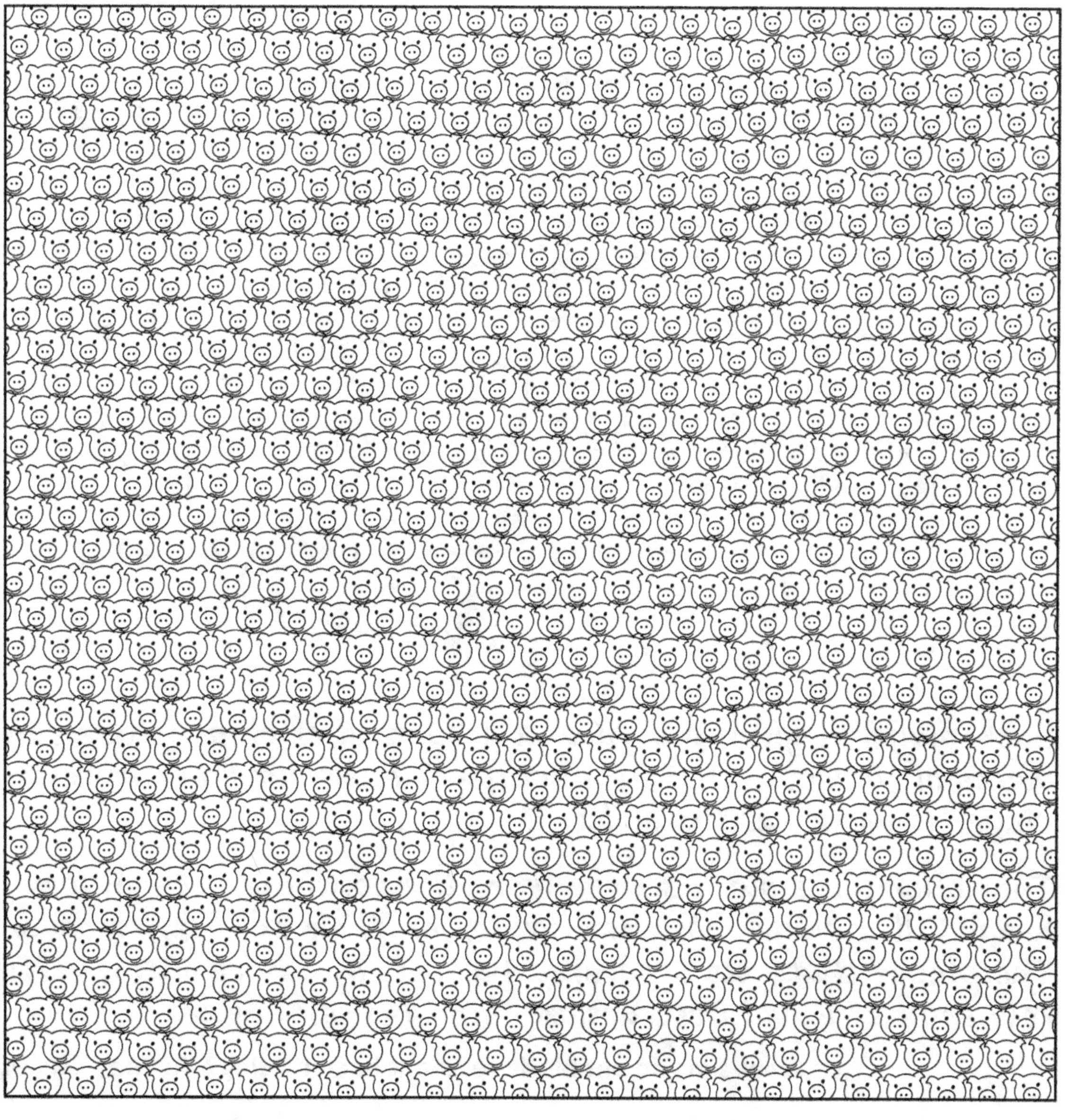

Rural Destruction Bingo

Surviving today's industrial farming practices takes more than skill, it takes luck.
Can you be a winner in the game of Rural Destruction?

Make a copy of the facing page and glue it to a piece of cardboard or foam-core board. Then cut the squares apart using scissors or a hobby knife. [Put the squares in a bucket or bowl.]

Next copy the Rural Destruction cards on the following pages, cut them out, and separate each pair.

ANTIBIOTIC RESISTANT BACTERIA
RESISTANT WEEDS
BOOM SPRAYER
DEPLETED AQUIFERS
CHEAP FOREIGN LABOR
LAND GRANT COLLEGES
ASTHMA
NUISANCE LAWS
HIGH LAND ASSESSMENTS
HOME FORE-CLOSURES
CARCASS PILES
MAGGOTS
GMO CROPS
FISH KILLS
MANURE SPILLS
GAS COUNTER
EROSION
SAME SPECIES FEEDING
MR. FACTORY FARM
REDUCED LIFE EXPECTANCY
REAL ESTATE DISCLOSURE FORM
POVERTY
BLUE BABIES
AG SUBSIDIES
BOVINE GROWTH HORMONE
CHICK GRINDER
DEC DOES NOTHING
MAXIMIZE THE PROFITS
SICK CHILDREN
MANURE LAGOONS
LOW HOME VALUES
WATER FILTER
PESTICIDES
HERBICIDE DRIFTS
VOLUNTARY GUIDE LINES
CITY-CENTRIC MEDIA
ALGAL BLOOMS
BUREAUCRATS
BLOW FLIES
BIG AG

Each player is given a card and a handful of chips [these can be pennies, pieces of paper, or "USDA Certified" labels cut from milk cartons.] One player is picked as the caller.

The caller picks out a square from the container at random and reads the hazard. If a player has that hazard on their card, they cover it with a chip. The player that gets 5 covered hazards in a horizontal, vertical or diagonal row shouts "Rural Destruction!" and is the winner.

Q: What song do antibiotic-resistant bacteria sing? A: *The Kill is Gone.*

DEPLETED AQUIFERS	HYDROGEN SULFIDE	MR. FACTORY FARM	TOXIC GAS PLUMES	BUREAUCRATS	BLOW FLIES	HOME FORE-CLOSURES
BEACHES CLOSED	AGAL BLOOMS	ANTIBIOTIC RESISTANT BACTERIA	BOVINE GROWTH HORMONE	DEC DOES NOTHING	MANURE LAGOONS	OFFAL PITS
PINK SLIME	DEAD ZONE	METHANE	PARTICLE MASKS	GAS COUNTER	STENCH WINDS	BLUE BABIES
HERBICIDE DRIFTS	PESTICIDES	MANURE SPILLS	FLOOD PLAIN	BIODIGESTER	USDA IN THEIR PAY	FISH KILLS
NUISANCE LAWS	EROSION	AG STUDY	POLITICIANS	SAME SPECIES FEEDING	CANCER	WATER FILTER
AG PAID EXPERTS	SICK CHILDREN	POLLUTED WELLS	BOTTLED WATER	GRAVE STONES	AG SUBSIDIES	BUG ZAPPER
MAXIMIZE THE PROFITS	POVERTY	REDUCED LIFE EXPECTANCY	BOOM SPRAYER	CORRUPTION	CHICK GRINDER	VOLUNTARY GUIDE LINES
BIG AG	CARCASS PILES	NO SWIMMING	CITY-CENTRIC MEDIA	REAL ESTATE DISCLOSURE FORM	AG RUNOFF	CHEAP FOREIGN LABOR
HIGH LAND ASSESSMENTS	ASTHMA	LAND GRANT COLLEGES	MAGGOTS	GMO CROPS	WINTER MANURE SPREADING	AG TAX CREDIT
RESISTANT WEEDS	SPRAYING LIQUID MANURE	LOW HOME VALUES	RURAL ROADS DESTROYED	STUCK INDOORS	PERSISTENT COUGH	DEPRESSION

RURAL

DEPLETED AQUIFERS	POLLUTED WELLS	HYDROGEN SULFIDE	DEAD ZONE	CORRUPTION
NO SWIMMING	MR. FACTORY FARM	NUISANCE LAWS	TOXIC GAS PLUMES	AG SUBSIDIES
AGAL BLOOMS	EROSION	ANTIBIOTIC RESISTANT BACTERIA	MANURE LAGOONS	POVERTY
REDUCED LIFE EXPECTANCY	OFFAL PITS	CANCER	FISH KILLS	HERBICIDE DRIFTS
ASTHMA	BLUE BABIES	BUREAUCRATS	BOVINE GROWTH HORMONE	PERSISTENT COUGH

DESTRUCTION

RURAL

GMO CROPS	SICK CHILDREN	AGAL BLOOMS	SAME SPECIES FEEDING	MANURE SPILLS
VOLUNTARY GUIDE LINES	BLOW FLIES	MR. FACTORY FARM	BOTTLED WATER	AG SUBSIDIES
BOOM SPRAYER	RURAL ROADS DESTROYED	REDUCED LIFE EXPECTANCY	CARCASS PILES	WINTER MANURE SPREADING
LOW HOME VALUES	DEPRESSION	CANCER	NUISANCE LAWS	FLOOD PLAIN
AG STUDY	POLLUTED WELLS	PARTICLE MASKS	METHANE	PINK SLIME

DESTRUCTION

RURAL

HOME FORE-CLOSURES	BLOW FLIES	STENCH WINDS	DEPLETED AQUIFERS	SICK CHILDREN
NUISANCE LAWS	AG PAID EXPERTS	BEACHES CLOSED	POLITICIANS	ANTIBIOTIC RESISTANT BACTERIA
WATER FILTER	DEAD ZONE	DEC DOES NOTHING	FLOOD PLAIN	GAS COUNTER
HIGH LAND ASSESSMENTS	PESTICIDES	HYDROGEN SULFIDE	BUG ZAPPER	LOW HOME VALUES
GMO CROPS	BLUE BABIES	BIODIGESTER	USDA IN THEIR PAY	GRAVE STONES

DESTRUCTION

RURAL

MANURE LAGOONS	MAXIMIZE THE PROFITS	ANTIBIOTIC RESISTANT BACTERIA	MAGGOTS	HOME FORE-CLOSURES
AG TAX CREDIT	TOXIC GAS PLUMES	CITY-CENTRIC MEDIA	LAND GRANT COLLEGES	BIG AG
BOVINE GROWTH HORMONE	CHICK GRINDER	PINK SLIME	REAL ESTATE DISCLOSURE FORM	HERBICIDE DRIFTS
BUREAUCRATS	CHEAP FOREIGN LABOR	AG RUNOFF	FISH KILLS	RESISTANT WEEDS
AGAL BLOOMS	SPRAYING LIQUID MANURE	MANURE SPILLS	STUCK INDOORS	CANCER

DESTRUCTION

RURAL

SICK CHILDREN	BIG AG	CORRUPTION	VOLUNTARY GUIDE LINES	WINTER MANURE SPREADING
SPRAYING LIQUID MANURE	MAGGOTS	CHICK GRINDER	SAME SPECIES FEEDING	BOTTLED WATER
FISH KILLS	RURAL ROADS DESTROYED	RESISTANT WEEDS	CITY-CENTRIC MEDIA	BOVINE GROWTH HORMONE
PERSISTENT COUGH	REAL ESTATE DISCLOSURE FORM	CARCASS PILES	NO SWIMMING	AG RUNOFF
CHEAP FOREIGN LABOR	NUISANCE LAWS	STUCK INDOORS	DEPRESSION	ANTIBIOTIC RESISTANT BACTERIA

DESTRUCTION

RURAL

AG TAX CREDIT	PESTICIDES	USDA IN THEIR PAY	STENCH WINDS	POLITICIANS
GRAVE STONES	BIODIGESTER	LAND GRANT COLLEGES	EROSION	BOOM SPRAYER
OFFAL PITS	BEACHES CLOSED	DEC DOES NOTHING	METHANE	GAS COUNTER
AG STUDY	PARTICLE MASKS	AG PAID EXPERTS	WATER FILTER	BUG ZAPPER
MAXIMIZE THE PROFITS	HIGH LAND ASSESSMENTS	POVERTY	TOXIC GAS PLUMES	ASTHMA

DESTRUCTION

RURAL

HERBICIDE DRIFTS	BLUE BABIES	PESTICIDES	MANURE SPILLS	VOLUNTARY GUIDE LINES
NO SWIMMING	AG PAID EXPERTS	NUISANCE LAWS	CANCER	GMO CROPS
REAL ESTATE DISCLOSURE FORM	FISH KILLS	MAXIMIZE THE PROFITS	FLOOD PLAIN	MAGGOTS
POLLUTED WELLS	POVERTY	STENCH WINDS	ASTHMA	CITY-CENTRIC MEDIA
CHICK GRINDER	AG SUBSIDIES	REDUCED LIFE EXPECTANCY	CORRUPTION	SICK CHILDREN

DESTRUCTION

RURAL

AGAL BLOOMS	SPRAYING LIQUID MANURE	LOW HOME VALUES	DEPRESSION	TOXIC GAS PLUMES
HOME FORE-CLOSURES	DEPLETED AQUIFERS	PERSISTENT COUGH	HYDROGEN SULFIDE	DEAD ZONE
AG TAX CREDIT	WINTER MANURE SPREADING	MR. FACTORY FARM	BLOW FLIES	ANTIBIOTIC RESISTANT BACTERIA
METHANE	NUISANCE LAWS	BOVINE GROWTH HORMONE	OFFAL PITS	PINK SLIME
DEC DOES NOTHING	HIGH LAND ASSESSMENTS	MANURE LAGOONS	FISH KILLS	CANCER

DESTRUCTION

I Can Open My Window

A Child's Poem

I can open my window
When there's no big machine
Spraying that brown stuff
On my toys that are clean.

I can open my window
When I don't see those flies
That land on my food
When I don't use my eyes.

I can open my window
When mom hangs out the clothes
When she peeks out the door
And don't wrinkle her nose.

I can open my window
And stick out my head
When they don't spray the herbcide
That makes the plants dead.

I can open my window
And turn on the fan
When the Bad is not out there
And my mom says I can.

Q; What's the government's answer to airborne factory farm pollution?
A: Buy your children respirators.

YOU'VE GOTTA BE KIDDING! IT'S NO JOKE!

Q: What are the last words a factory farm cow hears?
A: "She's down in milk production."

Q: What's brown and white and smells?
A: The side of your house when they're spraying manure.

Q: What steps did federal regulators take to meet their 2015 goal of reducing the Gulf of Mexico Dead Zone?
A: They pushed back the deadline to 2025.

Q: What's the similarity between a factory farm and a traditional farm?
A: The word "farm."

Q: Are factory farms always destroying the environment with manure spills?
A: No, they're doing it in a number of ways.

Q: What do you get if you cross a river with a manure spill?
A: Covered with manure.

Q: How do factory farms fix nutrient pollution spots?
A: With advertising spots.

Q: What's gray and yellow and red all over?
A: Chicks in a grinder.

MORTALITY MAZE

Find your way between the composting cows without crossing the seepage.

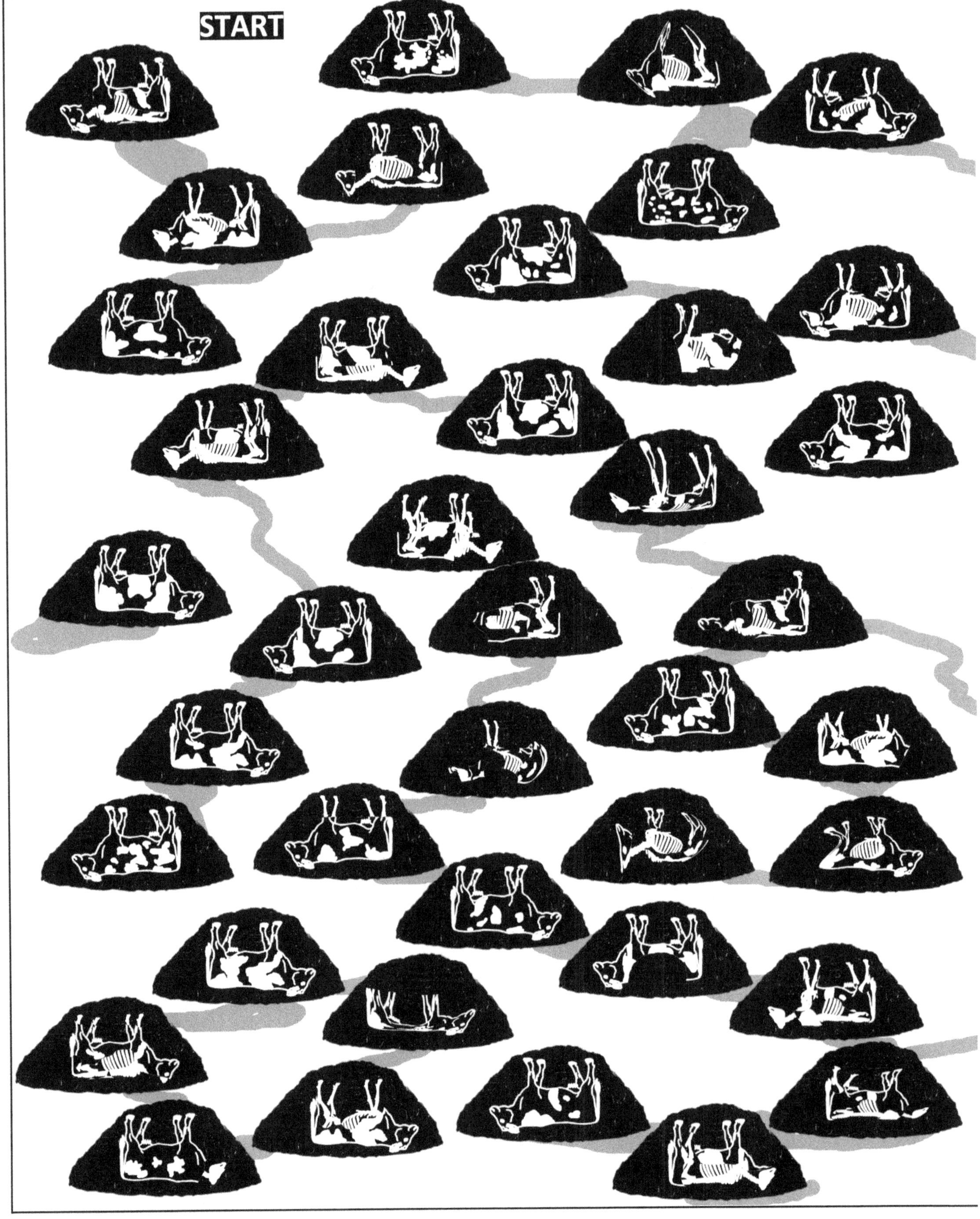
START

FINISH

When you have completed the maze, find the two identical cow carcasses.

FARM WORKER MORTALITY DISPOSAL

You can't *Feed the World* without making sacrifices, and it's not always your neighbors — worker death on corporate farms is just another natural occurrence, so go green and go under the radar with worker mortality composting and it's "problem solved."

Brief anonymous surveys reveal a widespread practice of improper mortality disposal. Bodies left to decay naturally above ground or buried in shallow pits pose risks to surface and groundwater and endanger the health of livestock.

Composting provides an inexpensive alternative for disposal of all dead workers. The practice does require space on your land to construct the compost piles and takes from two to six months for the body to decompose.

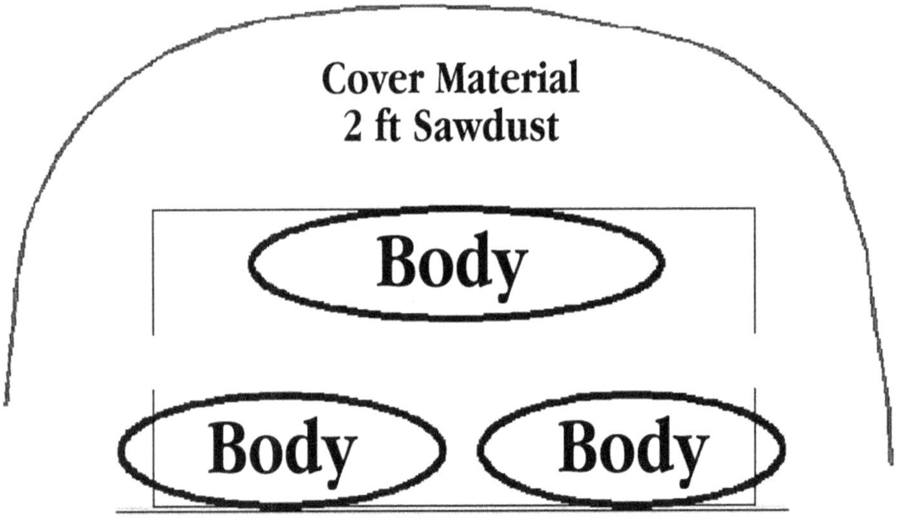
Cover Material
2 ft Sawdust
Body
Body
Body

Absorptive Material Layer
Wood Chips 18-24 inches

It is recommended to reuse finished compost as the base for the next pile. The remaining bones add structure to the base material for improved aeration. The composted material can also be used on hay, corn, winter wheat, tree plantations and forestland. Applying the compost to "table-top" crops directly consumed by people is not recommended at this time.

Tip: Attaching worker termination notices to the pile is an easy way to keep your records up to date.

Special Thanks to Contributors and Supporters

[illegible] - Director of [illegible]e Farm Services
[illegible] - [illegible] Corporate Science Teaching and Research Center
[illegible]y Agricultural Development Project
[illegible] University Corporate Extension
[illegible] Agriculture Council
Anonymous Sponsor Committee

A rich factory farmer named Fred . . .

A limerick is a form of verse, often humorous, with a five line rhyme scheme of AABBA, in which the first, second and fifth line rhyme, while the third and fourth lines are shorter and share a different rhyme. Here are a few examples:

A rich factory farmer named Fred
Had stuffed so many cows in a shed
He had bragged that, in fact
They were so closely packed,
They would stand whether living or dead.

Some geneticists hungry for fame
Used bad science to bolster their claim
And the bugs they created
Were a plague when they mated,
But they never accepted the blame.

An adamant GMO grower
Made his corporate costs even lower
With cheap foreign labor
Political favors
And clout to make slow studies, slower.

There was a fat hog on the farm
Who never did anyone harm
But with sisters and brothers
And 10,000 others
The neighbors looked on with alarm.

Now, try making up your own industrial farming limericks, using these lines as first lines:

The old Bed and Breakfast was closed
The DEC won't do a thing
Manure spilled once more today
The blow flies have gotten so bad
The offal bins just 'cross the street
A farmer who still wanted more
There was a hog farm in Nantucket

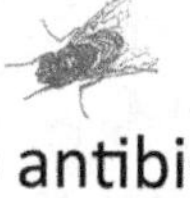

If a factory farm cow laughs really hard, do antibiotics come out of her nose?

FARM HARM
COLORING PAGE

COLOR WHAT THIS HOG FARM LOOKS LIKE AFTER A HURRICANE

[HINT: YOU WON'T NEED MANY COLORS]

FLOOD PLAIN HOG FARM

EASY MAZE
Herbicide Drift Maze
START
END

The babies need help, and it has to be swift,
Can you guide their parents through the herbicide drift?

I Wandered Lonely As A Cloud
(long after William Wordsworth)

I wandered lonely as a cloud
That floats on high o'er streams and hills,
When all at once I saw a crowd,
And trucks they use to clean up spills;
Upon the lake, beneath the trees,
The stench lay heavy in the breeze.

Continuous as the stars that shine
And twinkle in the Milky Way,
They march in never-ending line,
In darkened spill and bright decay;
Ten thousand dead fish met my glance,
In fly-filled pestilential dance.

And what I felt and want to say,
I cannot get outside of me,
Still hard and empty as that day,
Like plastic bottles on the sea;
I gazed—and gazed—in stone chipped thought,
What wealth and selfishness has brought.

And often when in bed I lie
In sleeplessness, and hard clenched mood,
That scene will flash upon my eye,
The death of Nature's plentitude;
And then my heart with anger fills
At those who kill our world with spills.

(With apologies to William Wordsworth, and none to industrial farming.)

Sustainable Truth Billboards

Wouldn't it be great to have honest advertising . . .

ONLY 1% OF AMERICANS ARE FARMERS
And 85% of America
is so far removed from Agriculture,
they do not understand what we do,
even when we tell them . . .
That's how we get away with it.

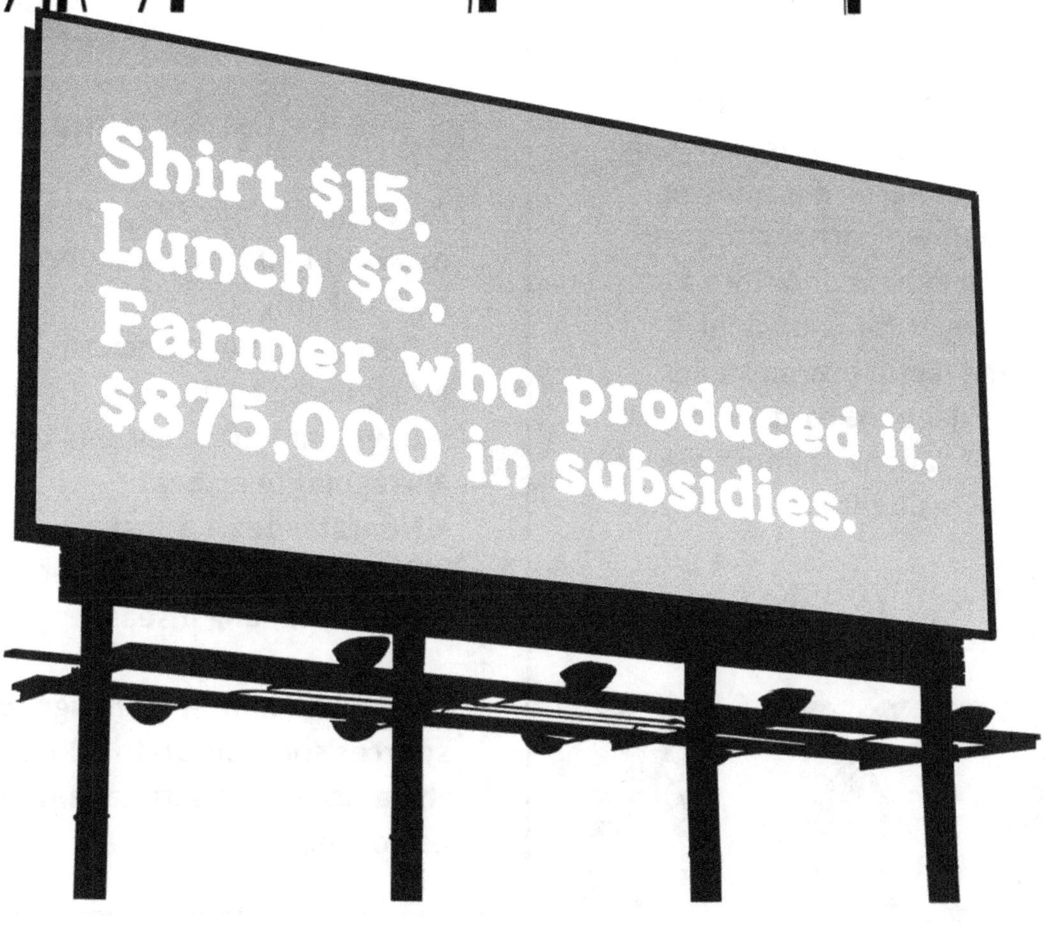
Shirt $15,
Lunch $8,
Farmer who produced it,
$875,000 in subsidies.

Child Safety Tips — *Respirators 101*

Your child is always at risk from toxic gases and infectious particles — even if they can't see or smell them.
How can you make sure they are safe?
Teach your children: **FARM HARM**

1 What do you mean by respirators and why do my children need them?

A respirator is a device to protect your child from inhaling the more than 500 VOCs, toxic gases and dangerous airborne pathogens produced by factory farms.

2 We don't live near a farm - aren't my children safe?

- **Scientific studies show that gas plumes and particulate drifts can travel for miles without dispersing.**
- **Manure and herbicide spraying, and windblown particulates can turn the field next door into a deadly danger for your child.**
- **Expanding factory farms build "satellite" lagoons to store millions of gallons of liquid manure near rural residences.**

3 Who can you go to when you need help?

Local Authorities | **DEC** | **USDA**

4 Where do airborne particles go?

Particle Size and Effect:

5.5 - 9.2 microns - Lodges in nose and throat

3.3 - 5.5 microns - Main breathing passages

2.0 - 3.3 microns - Small breathing passages

1.0 - 2.0 microns - Bronchi

0.3 - 1.0 microns - Air sacs

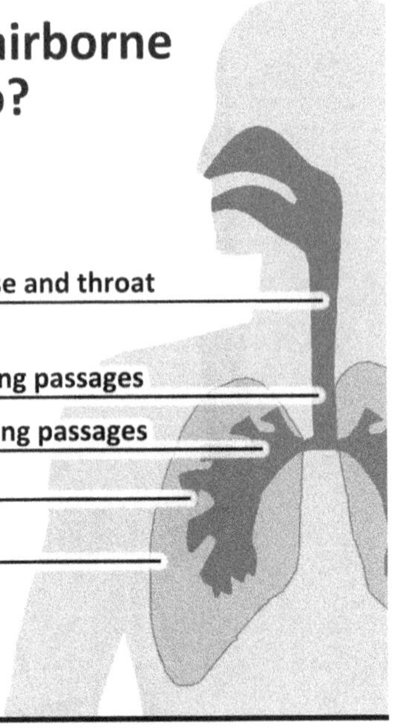

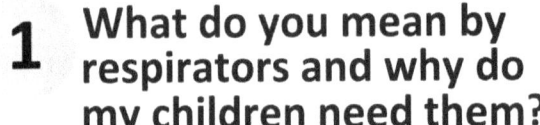

5 Is it dust, or dangerous?

According to the EPA, children exposed to particle pollution run a high risk of:

- **Decreased lung function**
- **Aggravated asthma**
- **Development of chronic bronchitis**
- **Irregular heartbeat**
- **Nonfatal heart attacks**
- **Premature death in children with heart and lung disease**

Particulate matter from agricultural sources contains up to 100 times the amount of bacteria and fungus as normal air.

7 Types of Respirators

Air-Purifying | **Atmosphere Supplying**

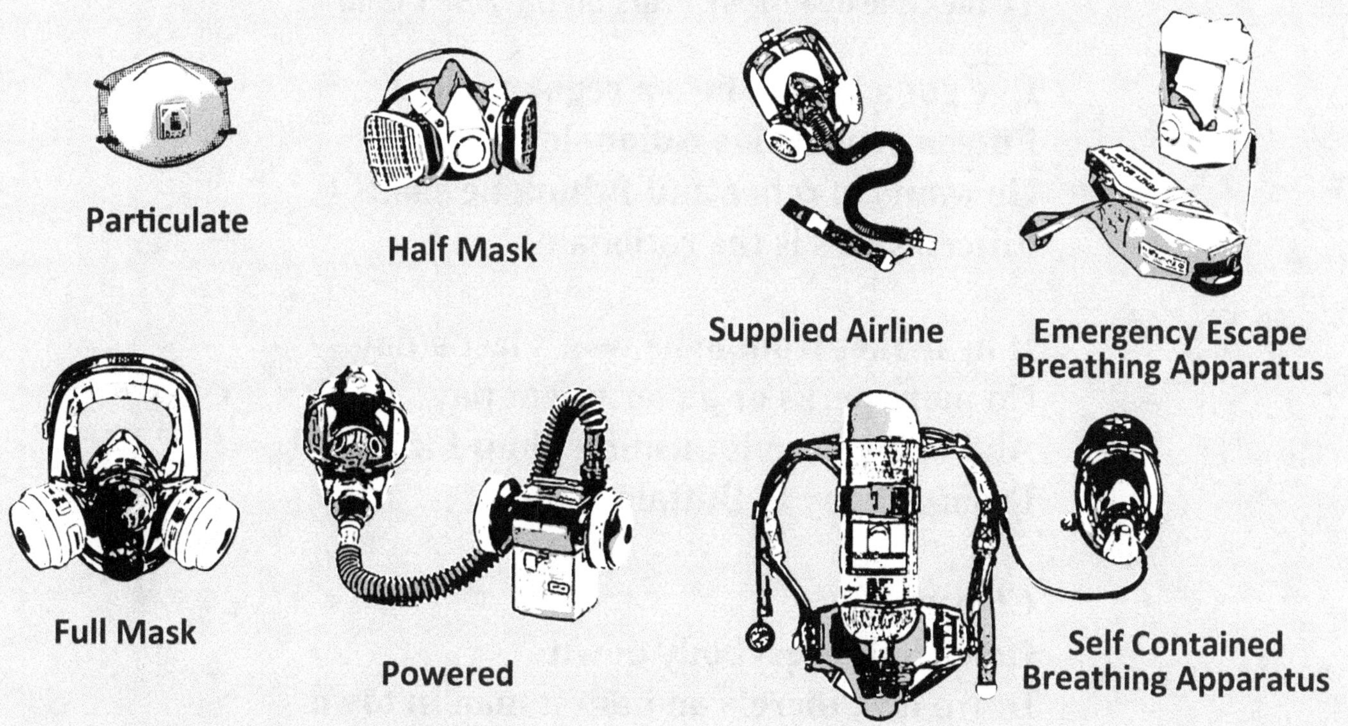
Particulate
Half Mask
Supplied Airline
Emergency Escape Breathing Apparatus
Full Mask
Powered
Self Contained Breathing Apparatus

6 It's not just "cow farts" - and it's not funny:

Among the more than 400 VOCs and toxic gases emitted by factory farms, rural children are exposed to:

- **Hydrogen Sulfide - prolonged exposure may cause convulsions, coma, brain and heart damage, even death.**
- **Methane - exposure may cause headaches, heart palpitations, cognitive disorders, dizziness, decrease in motor coordination, nausea, and suffocation.**
- **Ammonia - exposure to this gas may cause severe chest pain, difficulty in breathing, temporary or permanent blindness, dizziness, vomiting, and death.**

Even after decades of warnings and scientific papers citing the need for further research — *No comprehensive or long-term study of the effects of industrial farming emissions on the rural community has ever been made.*

8 What to teach your child:

Do not touch or walk in fields around a farm — or you may die.

Always wear a particle mask or respirator when playing outside.

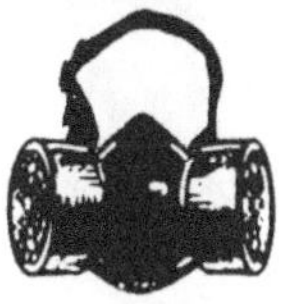

Stay away from manure lagoons — or you may die.

NO!

. . . . Parents . . . .

Agricultural studies are intended to protect the profits of farmers, not the health of your children. *Teach them about Farm Harm.*

"Down Low, Everybody Down"

To the tune of *Fifteen Years on the Erie Canal*

I've got a friend, he's a regulator pal
Fifteen years is his rationale
He wants to retire and I think he shall
Fifteen years is the rationale

If he makes trouble he won't last a day
No more perks or government pay
And every permit pushing pimp I know
From Albany to Buffalo

Chorus:
Down low, everybody down
Down low, there's an honest man in town
And you'll always get a favor
And you'll always know your pal
Cause a quid pro quo's the only bureaucratic rationale

There's my friend Al, today we're sure in luck
Fifteen years is his rationale
And when in Rome, all Romans take a buck
Fifteen years is the rationale

One more tip and back we go
A fistful of permits to spread and grow
And every permit pushing pimp I know
From Albany to Buffalo

Chorus:
Down low, everybody down
Down low, there's an honest man in town
And you'll always get a favor
And you'll always know your pal
Cause a quid pro quo's the only bureaucratic rationale

Factory Farm Feely Bag

"EEEWW! What is it?"

ground chick slurry

partially composted hog feet

ruminally inert fats

hydrolyzed poultry feather meal

offal bin scrapings

Cyanobacteria clumps

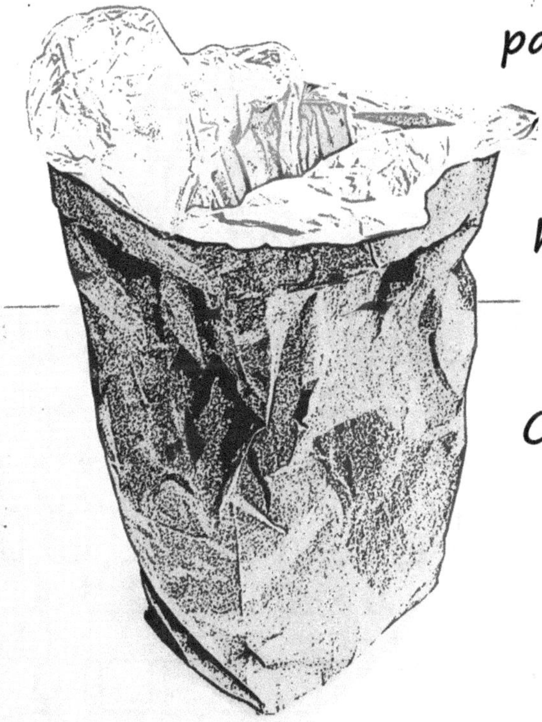

Take a plastic shopping bag and use it to line a brown paper bag. Then hide a different factory farm item in it each time, such as the ones listed above.

Ask the children to put their hand in the bag and guess what it is, without looking.

Consumers can choose whether or not they want hormones used — or can they?

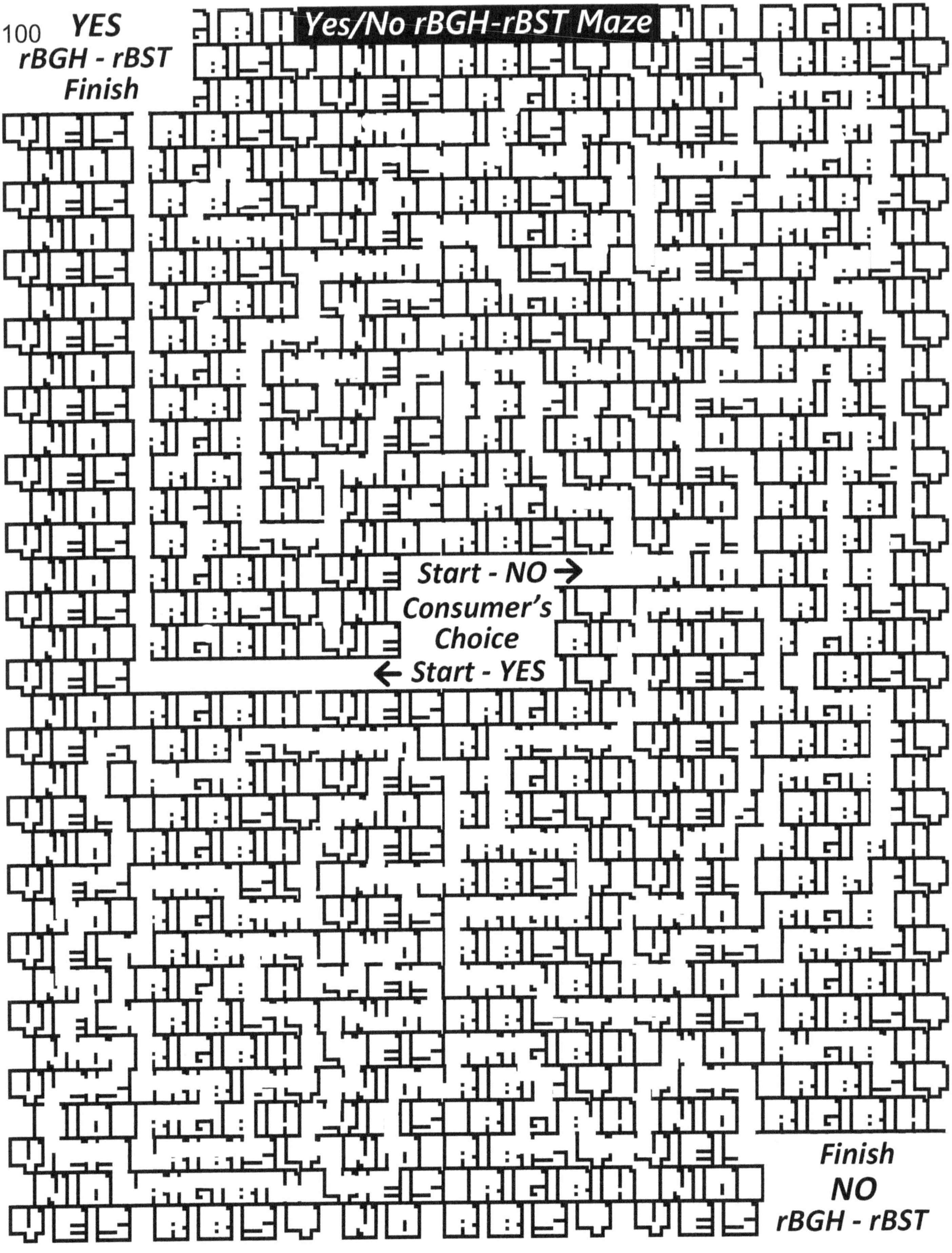
Yes/No rBGH-rBST Maze
YES
rBGH - rBST
Finish
Start - NO →
Consumer's
Choice
← Start - YES
Finish
NO
rBGH - rBST

The Never Ending Story

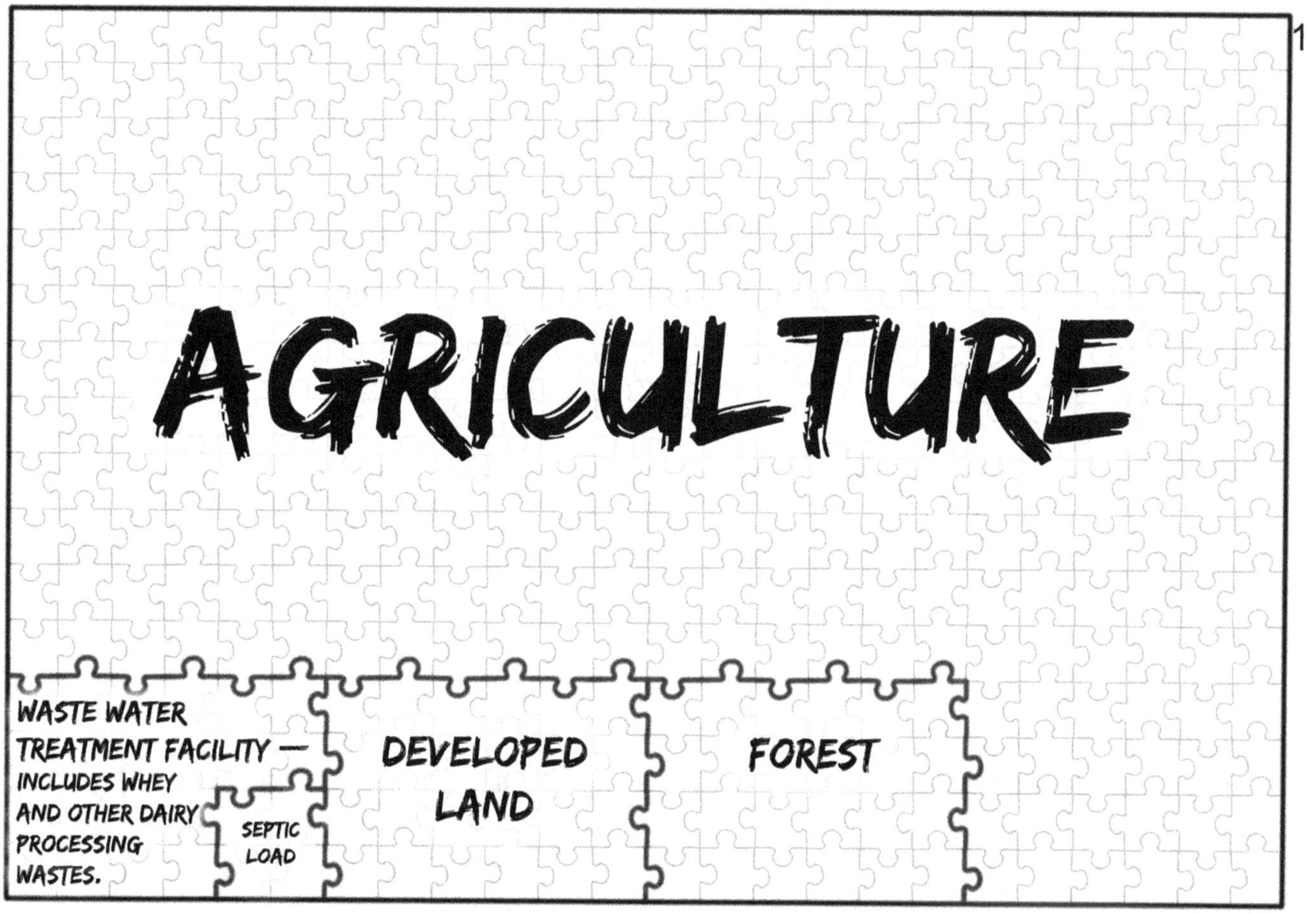
AGRICULTURE
WASTE WATER TREATMENT FACILITY — INCLUDES WHEY AND OTHER DAIRY PROCESSING WASTES.
SEPTIC LOAD
DEVELOPED LAND
FOREST

The Lake's Phosphorus Loading Puzzle
Now that it's pieced together, what's next?

Write a story that best predicts the future of the lake. The statistics and quotes are all factual. The lake is unnamed — the community is yours.

- Agriculture is responsible for 82% of the phosphorus loading in the lake.
- Local politicians and bureaucrats on farming: "They're as important as any lake."
- 3 out of the 5 voting members of the Soil and Water Conservation Committee are heavily involved in farming.
- State Harmful Algal Blooms Summit: "We don't know what's causing it."
- No regulations have so far been proposed to limit the agricultural phosphorus loading of the lake.

Q; At what point are residents removed from the decision making process? A: At the beginning.

Cracking the Ag Code

The Elephant in the Room

When your industry is responsible for *four times as much pollution as every other source combined . . .* What do you do?

Control the information

The Agriculture Lobby ensures that factory farm pollution is kept under the radar by controlling how it is presented to the public. One commonly used technique is to hide it in a group of sources and point to other culprits.

Agricultural Land Grant Colleges receive most of their funding from industry sources and are the first line of publicly disseminated information.

"Elevated nitrate levels in groundwater are usually caused by excess fertilizer applications to crops, lawns, gardens, parks, or golf courses; improperly handled livestock manures or intensive livestock production facilities; failing septic systems; decaying vegetation; of wastewater treatment plant discharge." — *Virginia Cooperative Extension*

01000011 01110010 01100001 01100011 01101011

"Sources of excess nitrate in water caused by human activities include fertilizers, on-site sewage systems (such as septic tanks and lagoons), wastewater treatment effluent, animal wastes, industrial wastes, and food processing." — *New Mexico State Cooperative Extension*

Land Grant Colleges will frequently target rural residents who are poor and have little power or media representation.

"Nitrogen not taken up by crops can leach through the soil to groundwater and then flow to recharge areas or private wells. Residents in rural communities typically use on-lot septic systems and some homeowners relay on lawn fertilizers. These too can be sources of nitrate in drinking water."
— *PennState Extension*

Agriculture is by far the biggest source of nitrates in rural drinking water and, in almost every case, the cause of nutrient poisoning in wells.

State governments release runoff "information" through departments tasked with environmental oversight and enforcement.

"Nonpoint source pollution is generally associated with human land-disturbing activities such as:

Urban development
Construction
Agriculture
Recreation
Silviculture
Mineral exploration"

— *Wyoming Department of Environmental Quality*

The Media is not adverse to running a little interference for agricultural interests.

"Dead zones, which disrupt fishing industries and threaten aquatic species, are caused by industrial and agricultural runoff." — *The Washington Post*

Environmental groups also present nutrient pollution in ways that minimize agriculture's liability.

"What Causes the Dead Zones? Heavy rains and melting snows washed massive amounts of nutrients—particularly nitrogen and phosphorous—from lawns, sewage treatment plants, farm land and other sources along the Mississippi River into the Gulf of Mexico." — *The Nature Conservancy*

00100000
01000011
01101111
01100100
01100101

The Iowa Farm Bureau manages to squeeze all of their state's industrial farming nutrient pollution between "golf courses" and "lawn treatment" fertilizers.

"Where do the increased nutrient levels cone from?
Physical Environment (Shape and flow of water body)
Fertilizers from Golf Courses, Agriculture, and Lawn Treatment
Atmospheric Nitrogen Deposits
Soil Nutrient Erosion/ Groundwater
Urban Run-off
Sewage Treatment Plant Discharge."

— *Iowa Farm Bureau*

In 2018, after 5 years of agriculture's voluntary *Nutrient Reduction Strategy*, Iowa's nutrient runoff has actually *increased*. Agriculture produces 90% of the state's nitrogen runoff.

From funding conflicting research, to controlling the committees that matter, "Big Ag" has created an information machine and regulatory environment that ensures their continued profits.

Take any Land Grant handout, runoff committee finding, CAFO regulatory announcement, or Ag study —

Can you spot the Elephant in the Room?

Factory Farm Bookshelf

"The best in industrial farming books, music, and games."

Family Fun Classics:

The Nutrient Plan Joke Book One good joke deserves another, and agricultural nutrient plans are about as useful "as a screen door in a submarine." Try out some of these nutrient plan jokes at your next barbecue [when the wind changes]:

Q: "How many nutrient plans does it take to screw in a light bulb?" A: "Forget it, they're useless."

Q: "What's brown and aqua and smells like bleach in a toilet?" A: "A nutrient spill in your swimming pool."

Q: "What did the nutrient plan say to the impaired lake?" A: "It's only a plan."

Q: "Why is a nutrient plan like a politician?" A: "They both make promises they know they won't fulfill."

Nutrient Plans are the only solution for agricultural pollution that factory farmers and politicians offer to the rural community. This 35 page book overflowing with derisive humor shows how much respect the rural community has for them.

Footprint Game The most successful industrial farms always make the most pollution, so if you want to make your mark in agriculture, make it a big one. The *Footprint Game* adheres to this principle by challenging competing players to make the biggest negative impact on the environment while focusing on the bottom line — money.

The Factory Farm Poop-up Book Teach your child toilet cleanliness with this washable vinyl book of factory farm "poop-up" scenes. Each scene portrays a different potty training cleanliness issue with admonitions in large print such as: *Antibiotic-resistant bacteria lagoon* — "Wash hands" and *Manure spill* — "Wipe yourself"
The book's colorful graphics explain both industrial farming *and* potty training in child-friendly terms. The "poop-ups" are fun and some are plain hilarious!

The Voluntary Guidelines Coloring Book A coloring book with an industrial farming twist — 50 scenes of natural beauty you can color however you want. Don't worry about staying within the guidelines, they're only voluntary.

DENIAL

A politician responds to media questions about a negative agricultural study:

A cooked-up concoction, a crude potpourri
And not at all what it is cracked up to be
Unwarranted wonderings, gratuitous muck
Dishonest, distorted, and coming unstuck
Assumed and ambiguous, iffy at least
Chimerical conclusions cunningly pieced
It's flimsy, fallacious, misleading, untrue
Unproven, uncalled for and unconfirmed too
Debatable, dubious, done to deceive
False, open to doubt, vague, and hard to believe
Illusive imposture, a shot in the dark
Duplicitous flimflam far wide of the mark
A groundless synthetic perfidious fake
Mock, foxy, slick, tricky, ma-li-ci-ous snake
Feigned, framed, made-up, slippery, mendacious sham
Disputable wicked-wrong rascally slam
Not at all, not either, nix, never, no
Now what was the question? It's a long time ago.

The End

Solutions

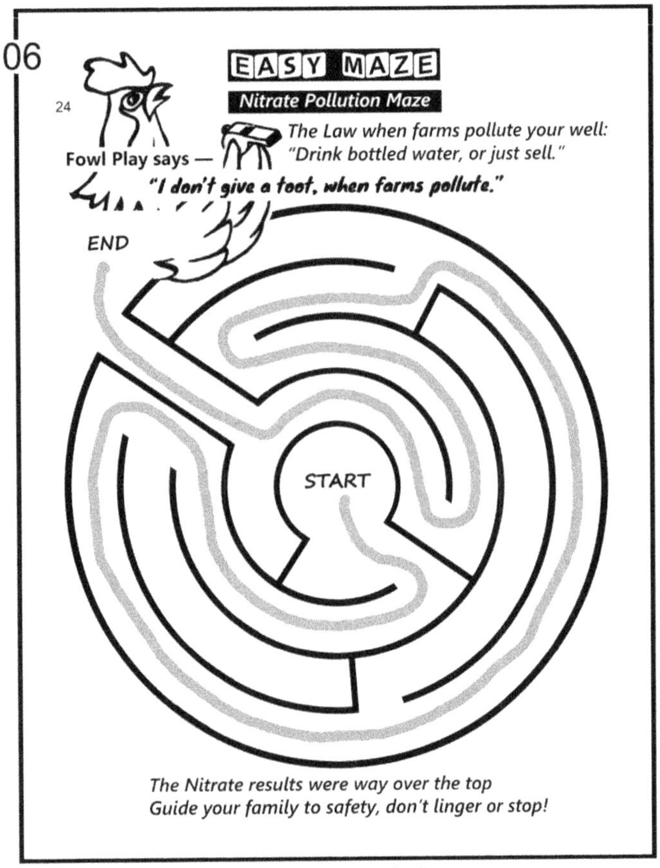
24
EASY MAZE
Nitrate Pollution Maze
The Law when farms pollute your well:
"Drink bottled water, or just sell."
Fowl Play says —
"I don't give a foot, when farms pollute."
END
START
The Nitrate results were way over the top
Guide your family to safety, don't linger or stop!

Page 24

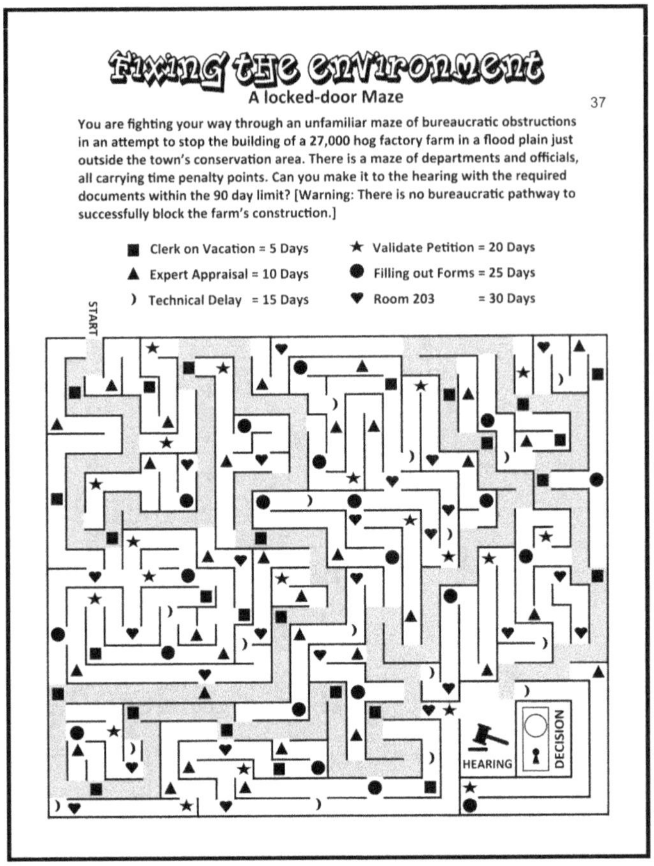
Fixing the environment
A locked-door Maze
37
You are fighting your way through an unfamiliar maze of bureaucratic obstructions in an attempt to stop the building of a 27,000 hog factory farm in a flood plain just outside the town's conservation area. There is a maze of departments and officials, all carrying time penalty points. Can you make it to the hearing with the required documents within the 90 day limit? [Warning: There is no bureaucratic pathway to successfully block the farm's construction.]
■ Clerk on Vacation = 5 Days
▲ Expert Appraisal = 10 Days
) Technical Delay = 15 Days
★ Validate Petition = 20 Days
● Filling out Forms = 25 Days
♥ Room 203 = 30 Days
START
HEARING
DECISION

Page 37

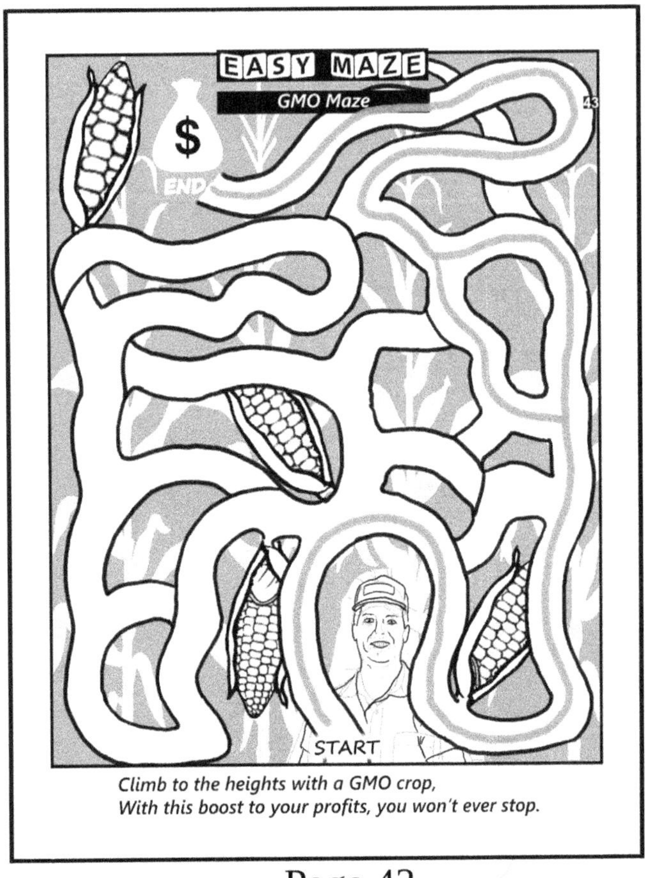
EASY MAZE
GMO Maze
43
$
END
START
Climb to the heights with a GMO crop,
With this boost to your profits, you won't ever stop.

Page 43

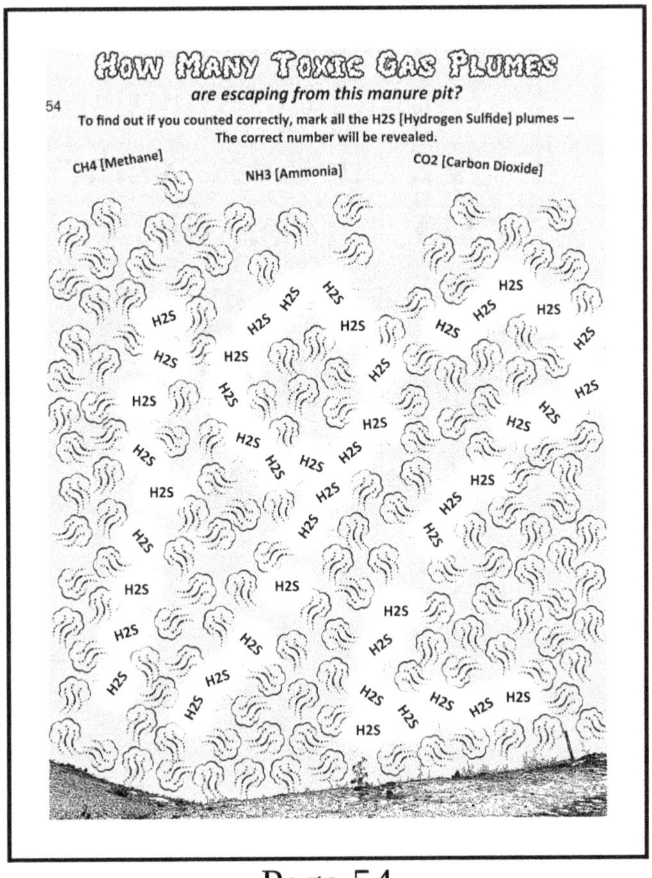
How Many Toxic Gas Plumes
are escaping from this manure pit?
54
To find out if you counted correctly, mark all the H2S [Hydrogen Sulfide] plumes — The correct number will be revealed.
CH4 [Methane]
NH3 [Ammonia]
CO2 [Carbon Dioxide]

Page 54

Solutions

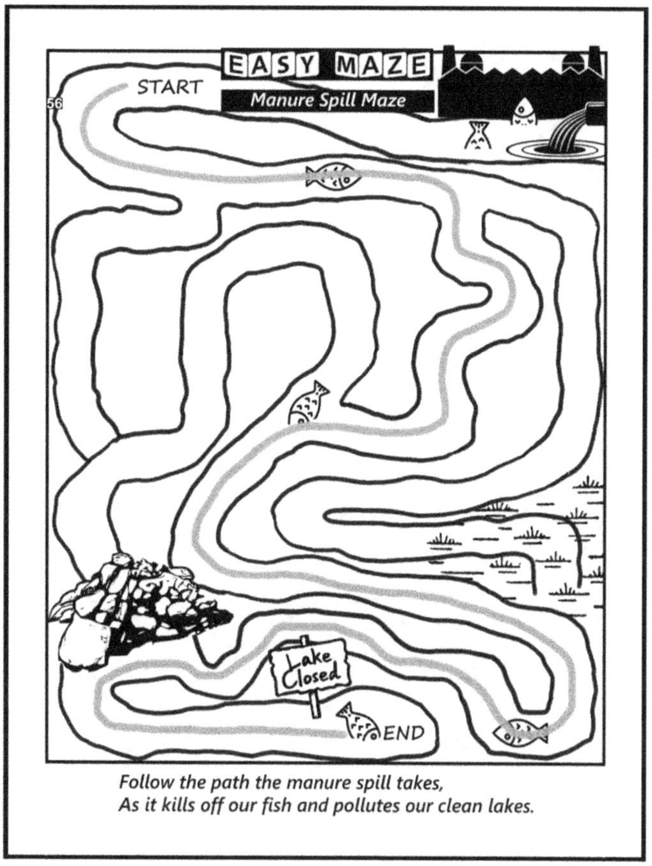
EASY MAZE
Manure Spill Maze
START
56
Lake Closed
END
Follow the path the manure spill takes,
As it kills off our fish and pollutes our clean lakes.

Page 56

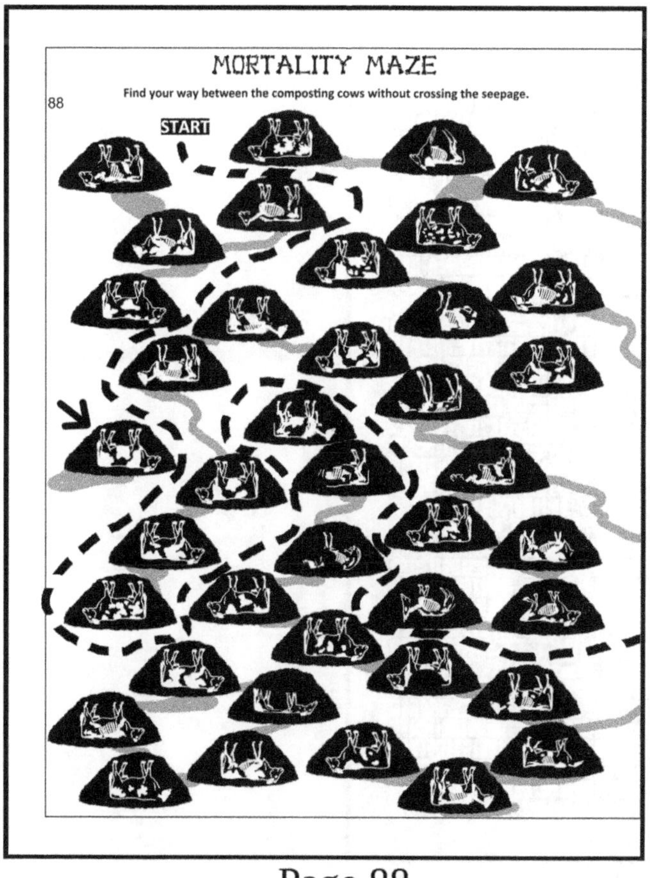
MORTALITY MAZE
Find your way between the composting cows without crossing the seepage.
88
START

Page 88

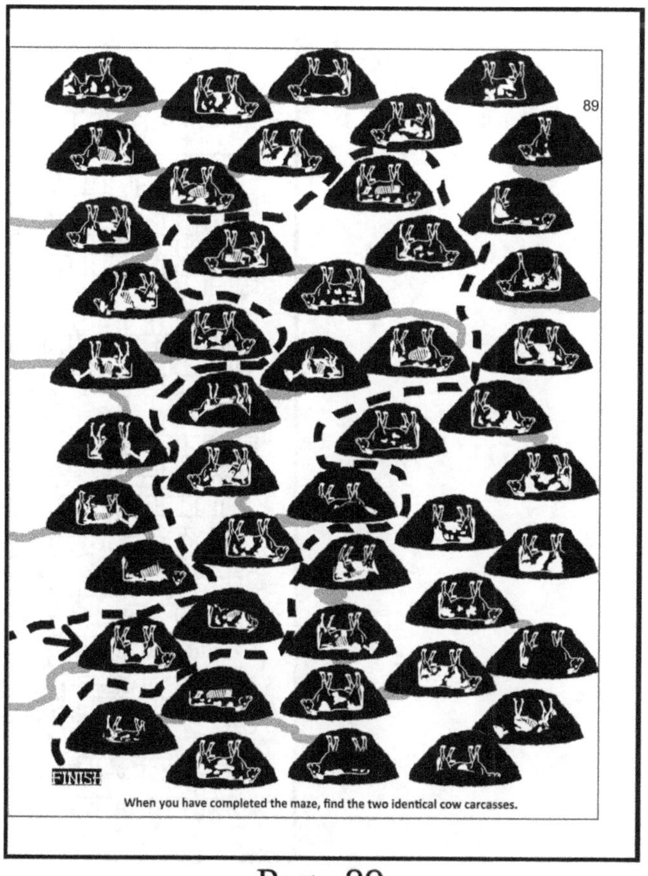
89
FINISH
When you have completed the maze, find the two identical cow carcasses.

Page 89

Solutions

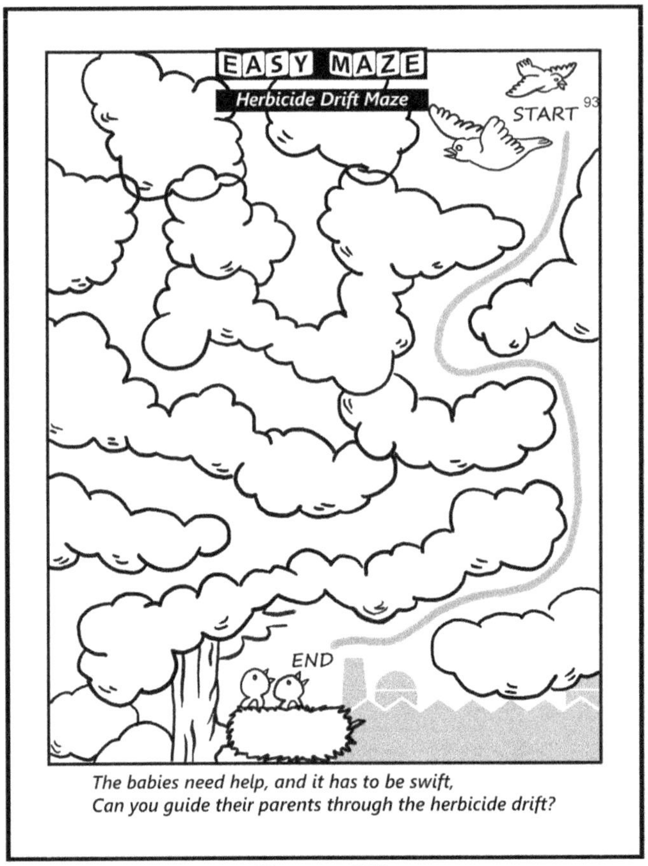
EASY MAZE
Herbicide Drift Maze
START
93
END
The babies need help, and it has to be swift,
Can you guide their parents through the herbicide drift?

Page 93

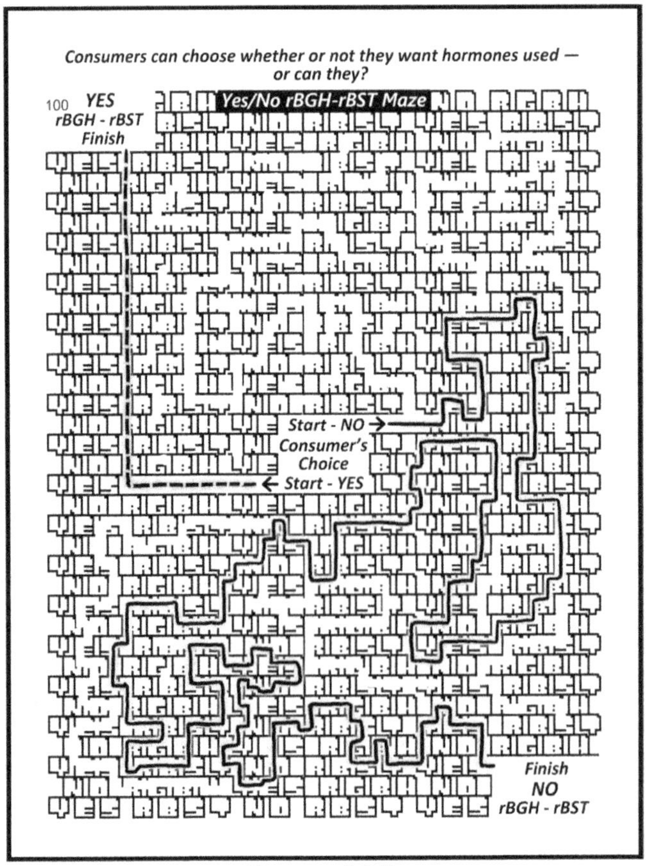
Consumers can choose whether or not they want hormones used — or can they?
100
YES
rBGH - rBST
Finish
Yes/No rBGH-rBST Maze
Start - NO
Consumer's
Choice
Start - YES
Finish
NO
rBGH - rBST

Page 100

About the Author

Doug Baird is an artist and writer living in Lansing, New York, who believes that both art and humor have transcendent properties.

His wide work experience has seen him travel within one corporation from the warehouse to corporate finance and product development, as well as in a variety of commercial art jobs, as a printer, illustrator, art director, and creative director.

Doug is project leader for the Idea Enhancement Project, a fiscally sponsored project of the New York Foundation for the Arts, exploring the use of art as a practical tool for increasing innovative and creative thinking. *IdeaEnhancement.org*

His blog, *Rural Tompkins County — The Road to Hell is Paved with Good Credentials*, investigates elitist policy making in New York, and its effect on the rural community.

He is the author of two poetry collections: *As a Poet, I have a Confession* and *Please Take Care when You Utter a Curse*, and a recently published picture book: *You Know You Live near a Factory Farm When Your Kids Go Fishing with a Pool Skimmer*.

Representative artworks can be viewed at *DougBairdArt.com*

Doug does not use cell phones or social media in order to spend more time on creative projects, quality of life, and pubs.

You Know You Live near a Factory Farm
When Your Kids Go Fishing with a Pool Skimmer

You Know You Live near a Factory Farm
When Your Kids Go Fishing with a Pool Skimmer
An Uncensored Look at Today's Farming Practices
by Doug Baird
This work is dedicated to the public domain
2018

#1 in the Factory Farm series

Available from online retailers world-wide

www.ingramcontent.com/pod-product-compliance
Lightning Source LLC
Chambersburg PA
CBHW060425010526
44118CB00017B/2361